Ford Capri II Owners Workshop Manual

by J H Haynes
Member of the Guild of Motoring Writers
and I M Coomber

Models covered:
Ford Capri II 1300 1298 cc ohv

Does not cover ohc engines

ISBN 0 85696 593 6

© Haynes Publishing Group 1977, 1981, 1988

ABCDE
FGHIJ
K

All rights reserved. No part of this book may be reproduced or transmitted in any form or by any means, electronic or mechanical, including photocopying, recording or by any information storage or retrieval system, without permission in writing from the copyright holder.

Printed in England   (338--7N2)

Haynes Publishing Group
Sparkford Nr Yeovil
Somerset BA22 7JJ England

Haynes Publications, Inc
861 Lawrence Drive
Newbury Park
California 91320 USA

Acknowledgements

Special thanks are due to the Ford Motor Company for the supply of technical information and certain illustrations. Castrol Limited provided lubrication data and the Champion Sparking Plug Company provided the spark plug illustrations. The bodywork repair photographs used in this manual were provided by Holt Lloyd Ltd. who supply 'Turtle Wax', 'Dupli-color Holts', and other Holts range products.

The Section of Chapter 10 dealing with the suppression of radio interference, was originated by Mr I. P. Davey, and was first published in *Motor* magazine.

Lastly, thanks are due to all of those people at Sparkford who helped in the production of this manual.

About this manual

Its aim

The aim of this manual is to help you get the best value from your car. It can do so in several ways. It can help you decide what work must be done (even should you choose to get it done by a garage), provide information on routine maintenance and servicing, and give a logical course of action and diagnosis when random faults occur. However, it is hoped that you will use the manual by tackling the work yourself. On simpler jobs it may even be quicker than booking the car into a garage and going there twice to leave and collect it. Perhaps most important, a lot of money can be saved by avoiding the costs the garage must charge to cover its labour and overheads.

The manual has drawings and descriptions to show the function of the various components so that their layout can be understood. Then the tasks are described and photographed in a step-by-step sequence so that even a novice can do the work.

Its arrangement

The manual is divided into thirteen Chapters, each covering a logical sub-division of the vehicle. The Chapters are each divided into Sections, numbered with single figures, eg 5; and the Sections into paragraphs (or sub-sections) with decimal numbers following on from the Section they are in, eg 5.1, 5.2, 5.3 etc.

It is freely illustrated, especially in those parts where there is a detailed sequence of operations to be carried out. There are two forms of illustration; figures and photographs. The figures are numbered in sequence with decimal numbers, according to their position in the Chapter — eg Fig. 6.4 is the fourth drawing/illustration in Chapter 6. Photographs carry the same number (either individually or in related groups) as the Section or sub-section to which they relate.

There is an alphabetical index at the back of the manual as well as a contents list at the front. Each Chapter is also preceded by its own individual contents list.

References to the 'left' or 'right' of the vehicle are in the sense of a person in the driver's seat facing forwards.

Unless otherwise stated, nuts and bolts are removed by turning anti-clockwise, and tightened by turning clockwise.

Vehicle manufacturers continually make changes to specifications and recommendations, and these, when notified, are incorporated into our manuals at the earliest opportunity.

Whilst every care is taken to ensure that the information in this manual is correct, no liability can be accepted by the authors or publishers for loss, damage or injury caused by any errors in, or omissions from, the information given.

Contents

	Page
Acknowledgements	2
About this manual	2
Introduction to the Capri II	4
Buying spare parts and vehicle identification numbers	6
Routine maintenance	7
Jacking and towing	9
Recommended lubricants and fluids	10
Tools and working facilities	11
Chapter 1 Engine	13
Chapter 2 Cooling system	40
Chapter 3 Carburation; fuel and exhaust systems	46
Chapter 4 Ignition system	57
Chapter 5 Clutch	67
Chapter 6 Gearbox	71
Chapter 7 Propeller shaft	84
Chapter 8 Rear axle	86
Chapter 9 Braking system	94
Chapter 10 Electrical system	106
Chapter 11 Suspension and steering	139
Chapter 12 Bodywork and fittings	150
Chapter 13 Supplement: Revisions and information on later models	167
Conversion factors	181
Safety first!	182
Index	183

Introduction to the Capri II

The Capri II models were first introduced in the United Kingdom in February 1974, the 1.3 litre ohv version being the smallest in the range.

The 1298 cc engine is basically the same as used on other Ford models and the mechanical layout is quite conventional.

Drive from the overhead valve engine is transmitted to the rear axle via a four-speed, all synchromesh gearbox and single piece propeller shaft.

The front suspension is independent with MacPherson struts and an anti-roll bar. The rear suspension has semi-elliptic leaf springs with telescopic shock absorbers and an anti-roll bar.

The three door Coupe body is of all steel welded integral construction.

Although the UK version is only 1 inch longer and 2¼ inches wider than the previous model Capri, the appearance of a larger car is obtained by the sleeker lines which evolved with the restyling. This is even more apparent from the inside due to the increase in load space and opening tailgate giving easier access to the luggage compartment.

A variety of optional extras is available, depending on the particular model and intended market.

1975 Ford Capri II 1.3

1979 Ford Capri II 1.3

Buying spare parts and vehicle identification numbers

Buying spare parts

Spare parts are available from many sources, for example: Ford garages, other garages and accessory shops, and motor factors. Our advice regarding spare part sources is as follows:

Officially appointed Ford garages - This is the best source of parts which are peculiar to your car and are otherwise not generally available (eg. complete cylinder heads, internal gearbox components, badges interior trim etc). It is also the only place at which you should buy parts if your car is still under warranty - non-Ford components may invalidate the warranty. To be sure of obtaining the correct parts it will always be necessary to give the storeman your car's vehicle identification number, and if possible, to take the 'old' part along for positive identification. Remember that many parts are available on a factory exchange scheme - any parts returned should always be clean! It obviously makes good sense to go straight to the specialists on your car for this type of part for they are best equipped to supply you.

Other garages and accessory shops - These are often very good places to buy materials and components needed for the maintenance of your car (eg. oil filters, spark plugs, bulbs, fanbelts, oils and greases, touch-up paint, filler paste, etc). They also sell general accessories, usually have convenient opening hours, charge lower prices and can often be found not far from home.

Motor factors - Good factors will stock all of the more important components which wear out relatively quickly (eg. clutch components, pistons, valves, exhaust systems, brake cylinders/pipes/hoses/seals/shoes and pads etc). Motor factors will often provide new or reconditioned components on a part exchange basis - this can save a considerable amount of money.

Vehicle identification numbers

Although many individual parts, and in some cases sub-assemblies, fit a number of different models it is dangerous to assume that just because they look the same, they are the same. Differences are not always easy to detect except by serial numbers. Make sure therefore, that the appropriate identity number for the model or sub-assembly is known and quoted when a spare part is ordered.

The vehicle identification plate is mounted on the right-hand front wing (fender) apron or on the bonnet locking platform, and may be seen once the bonnet is open. Record the numbers from your car on the blank spaces of the accompanying illustration. You can then take the manual with you when buying parts; also the exploded drawings throughout the manual can be used to point out and identify the components required. In certain instances, reference is included in the text of this manual to a build code. The code letters are the last two of a six letter grouping in the Vehicle No. section of the vehicle identification plate.

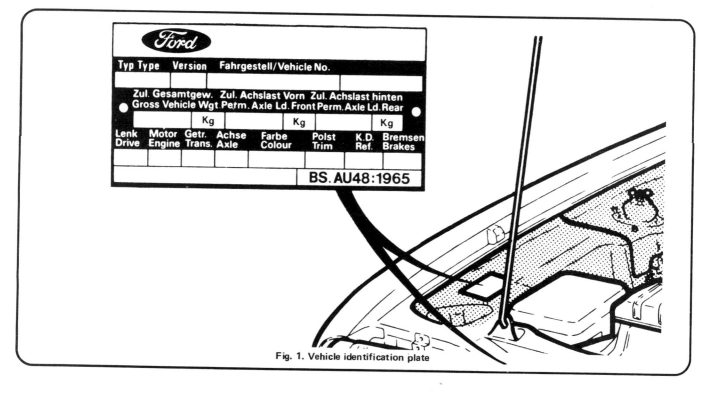

Fig. 1. Vehicle identification plate

Routine maintenance

Maintenance is essential for ensuring safety, and desirable for the purpose of getting the best in terms of performance and economy from your car. Over the years the need for periodic lubrication - oiling, greasing and so on - has been drastically reduced, if not totally eliminated. This has unfortunately tended to lead some owners to think that because no such action is required, components either no longer exist, or will last forever. This is a serious delusion. It follows therefore that the largest initial element of maintenance is visual examination and a general sense of awareness. This may lead to repairs or renewals, but should help to avoid roadside breakdowns.

Although the maintenance instructions listed below are basically those recommended by the manufacturer, they are supplemented by additional service checks which, through practical experience, the author recommends should be carried out at the intervals suggested.

Every 250 miles (400 km), weekly or before a long journey

Steering
  Check tyre pressures (when cold), including the spare.
  Examine tyres for wear and damage.
  Check steering for smooth and accurate operation.

Brakes
  Check reservoir fluid level. If this has fallen noticeably, check for fluid leakage (photo).
  Check for satisfactory brake operation.

Lights, wipers, horns, instruments
  Check operation of all lights.
  Check operation of windscreen wipers and washers.
  Check that the horn operates.
  Check that all instruments and gauges are operating.

Engine compartment
  Check engine oil level; top-up if necessary (photo).
  Check radiator coolant level.
  Check battery electrolyte level.

At first 3000 miles (5000 km), and for vehicles which operate under continuous stop/start conditions every subsequent 3000 miles (5000 km)

  Renew engine oil.
  Renew engine oil filter at first 3000 miles (5000 km) (photo).

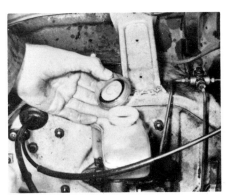
Brake fluid reservoir - typical

Topping-up engine oil

The engine oil filter location

Check the disc brake pads for wear

Check the rear brake linings for wear

Every 6000 miles (10000 km) or 6 months, whichever occurs first

Renew engine oil and oil filter.
Clean distributor points and reset gap.
Lubricate distributor.
Clean spark plugs and reset gaps.
Clean all HT leads and top of ignition coil.
Check ignition timing.
Check valve clearances.
Check tightness of inlet and exhaust manifold bolts.
Check condition of exhaust system.
Check condition and tension of the fanbelt.
Lubricate accelerator linkage.
Adjust engine idle speeds and mixture.
Examine cooling system hoses and check for leaks.
Clean/tighten battery terminals.
Check gearbox oil level.
Check rear axle oil level.
Check and top up if necessary the brake fluid reservoir.
Check clutch adjustment.
Check front brake pads for wear (photo).
Check rear brake linings for wear (photo).
Examine brake hoses for leaks and chafing.
Check handbrake adjustment.
Check steering linkage for wear and damage, and condition of ball joint covers.
Check front suspension linkage for wear and damage.
Check front wheel toe-in.
Check operation of all doors, catches and hinges. Lubricate as necessary.
Check condition of seatbelts and operation of buckles and inertia reels.
Check the tyre pressures and adjust if required. Inspect the general condition of tyre walls and treads.
Check and top up the windscreen washer reservoirs.
Check that all lights and indicators are in working order.
Check the tightness of the alternator mounting bolts.

Every 12000 miles (20000 km) or 12 months, whichever occurs first

Renew the contact breaker points.
Renew spark plugs.
Check condition of distributor cap and rotor.
Clean the positive crankcase ventilation (PCV) system.

Every 18000 miles (30000 km) or 18 months, whichever occurs first

Check tightness of rear spring mountings.
Change the air cleaner element.
Check and tighten if required, the rear spring U-bolts to the correct torque.

Every 24000 miles (40000 km) or 2 years, whichever occurs first

Dismantle, lubricate and adjust front wheel bearings.
Renew all rubber seals and hoses in braking system. Renew brake fluid.
Drain engine coolant. Renew antifreeze or inhibitor coolant mixture.

Every 36000 miles (60000 km)

Clean the front wheel bearings, repack with grease and adjust.
Drain and renew the hydraulic fluid - replace seals and hoses as required.

Jacking and towing

Jacking points

To change a wheel in an emergency, use the jack supplied with the vehicle. Ensure that the roadwheel nuts are released before jacking up the car and make sure that the arm of the jack is fully engaged with the body bracket and that the base of the jack is standing on a firm surface.

The jack supplied with the vehicle is not suitable for use when raising the vehicle for maintenance or repair operations. For this work, use a trolley, hydraulic or screw type jack located under the front crossmember, bodyframe side-members or rear axle casing, as illustrated. Always supplement the jack with axle stands or blocks before crawling beneath the car.

Towing points

If your vehicle is being towed, make sure that the tow rope is attached to the front crossmember or, if the vehicle is so equipped, by the front towing eye on the crossmember. If you are towing another vehicle, attach the tow rope to the lower shock absorber mounting bracket beneath the axle tube or, if the vehicle is so equipped, by the rear towing eye beneath the bumper.

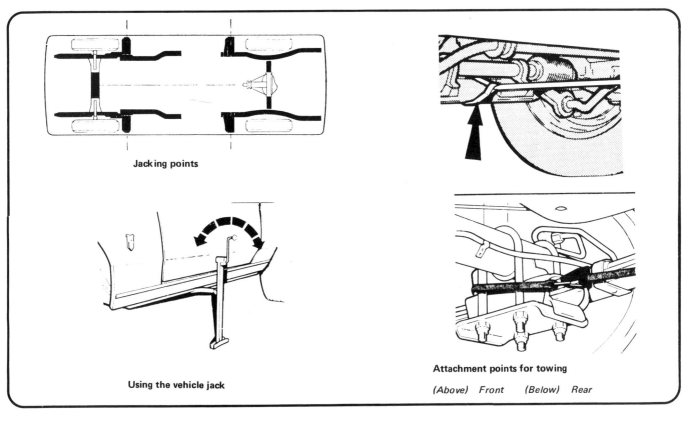

Jacking points

Using the vehicle jack

Attachment points for towing

(Above) Front (Below) Rear

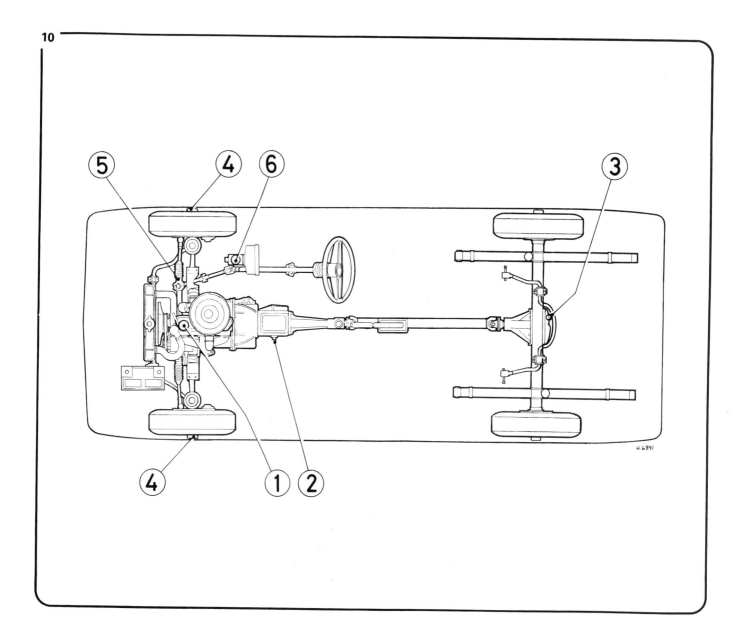

Recommended lubricants and fluids

Component or system	Lubrication type or specification	Castrol product
Engine (1)	Multigrade engine oil	Castrol GTX
Gearbox (2)	SAE 80EP gear oil	Castrol Hypoy Light
Rear axle (3)	SAE 90EP gear oil	Castrol Hypoy B
Front wheel bearings (4)	Multi-purpose lithium based grease	Castrol LM grease
Steering gear (5)	SAE 90EP gear oil	Castrol Hypoy B
Brake master cylinder (6)	SAE J1703, DOT 3 or 4, SAM-6C 9101-A (Amber) or ESEAM-6C 1001-A (Green)	Castrol Girling Universal Brake and Clutch fluid

Note: *The above are general recommendations only. Lubrication requirements vary from territory-to-territory and depend on vehicle usage. If in doubt, consult the operator's handbook supplied with the vehicle, or your nearest dealer.*

Tools and working facilities

Introduction

A selection of good tools is a fundamental requirement for anyone contemplating the maintenance and repair of a motor vehicle. For the owner who does not possess any, their purchase will prove a considerable expense, offsetting some of the savings made by doing-it-yourself. However, provided that the tools purchased are of good quality, they will last for many years and prove an extremely worthwhile investment.

To help the average owner to decide which tools are needed to carry out the various tasks detailed in this manual, we have compiled three lists of tools under the following headings: Maintenance and minor repair; Repair and overhaul; and Special. The newcomer is practical mechanics should start off with the 'Maintenance and minor repair' tool kit and confine himself to the simpler jobs around the vehicle. Then, as his confidence and experience grows, he can undertake more difficult tasks, buying extra tools as, and when, they are needed. In this way, a 'Maintenance and minor repair' tool kit can be built-up into a 'Repair and overhaul' tool kit over a considerable period of time without any major cash outlays. The experienced do-it-yourselfer will have a tool kit good enough for most repair and overhaul procedures and will add tools from the 'Special' category when he feels the expense is justified by the amount of use these tools will be put to.

It is obviously not possible to cover the subject of tools fully here. For those who wish to learn more about tools and their use there is a book entitled 'How to Choose and Use Car Tools' available from the publishers of this manual.

Maintenance and minor repair tool kit

The tools given in this list should be considered as a minimum requirement if routine maintenance, servicing and minor repair operations are to be undertaken. We recommend the purchase of combination spanners (ring one end, open-ended the other); although more expensive than open ended ones, they do give the advantages of both types of spanner.

Combination spanners - 7/16, 1/2, 9/16, 5/8, 11/16, 3/4 in AF
Combination spanners - 10, 11, 13, 14, 17 mm
Adjustable spanner - 9 inch
Engine sump/gearbox/rear axle drain plug key (where applicable)
Spark plug spanner (with rubber insert)
Spark plug gap adjustment tool
Set of feeler gauges
Brake adjuster spanner (where applicable)
Brake bleed nipple spanner
Screwdriver - 4 in. long x ¼ in. dia. (plain)
Screwdriver - 4 in. long x ¼ in. dia. (crosshead)
Combination pliers - 6 in.
Hacksaw, junior
Tyre pump
Tyre pressure gauge
Grease gun (where applicable)
Oil can
Fine emery cloth (1 sheet)
Wire brush (small)
Funnel (medium size)

Repair and overhaul tool kit

These tools are virtually essential for anyone undertaking any major repairs to a motor vehicle, and are additional to those given in the Basic list. Included in this list is a comprehensive set of sockets. Although these are expensive they will be found invaluable as they are so versatile - particularly if various drives are included in the set. We recommend the ½ in. square-drive type, as this can be used with most proprietary torque wrenches. If you cannot afford a socket set, even bought piecemeal, then inexpensive tubular box spanners are a useful alternative.

The tools in this list will occasionally need to be supplemented by tools from the Special list.

Sockets (or box spanners) to cover range in previous list
Reversible ratchet drive (for use with sockets)
Extension piece, 10 inch (for use with sockets)
Universal joint (for use with sockets)
Torque wrench (for use with sockets)
'Mole' wrench - 8 inch
Ball pein hammer
Soft-faced hammer, plastic or rubber
Screwdriver - 6 in. long x 5/16 in. dia. (plain)
Screwdriver - 2 in long x 5/16 in. square (plain)
Screwdriver - 1½ in. long x ¼ in. dia. (crosshead)
Screwdriver - 3 in. long x 1/8 in. dia. (electricians)
Pliers - electricians side cutters
Pliers - needle nosed
Pliers - circlip (internal and external)
Cold chisel - ½ inch
Scriber (this can be made by grinding the end of a broken hacksaw blade)
Scraper (this can be made by flattening and sharpening one end of a piece of copper pipe)
Centre punch
Pin punch
Hacksaw
Valve grinding tool
Steel rule/straight edge
Allen keys
Selection of files
Wire brush (large)
Axle stands
Jack (strong scissor or hydraulic type)

Tools and working facilities

Special tools

The tools in this list are those which are not used regularly, are expensive to buy, or which need to be used in accordance with their manufacturers instructions. Unless relatively difficult mechanical jobs are undertaken frequently, it will not be economic to buy many of these tools. Where this is the case, you could consider clubbing together with friends (or a motorists club) to make a joint purchase, or borrowing the tools against a deposit from a local garage or tool hire specialist.

The following list contains only those tools and instruments freely available to the public, and not those special tools produced by the vehicle manufacturer specifically for its dealer network. You will find occasional references to these manufacturers special tools in the text of this manual. Generally, an alternative method of doing the job without the vehicle manufacturers special tool is given. However, sometimes there is no alternative to using them. Where this is the case and the relevant tool cannot be bought or borrowed you will have to entrust the work to a franchised garage.

Valve spring compressor
Piston ring compressor
Ball joint separator
Universal hub/bearing puller
Impact screwdriver
Micrometer and/or vernier gauge
Carburettor flow balancing device (where applicable)
Dial gauge
Stroboscopic timing light
Dwell angle meter/tachometer
Universal electrical multi-meter
Cylinder compression gauge
Lifting tackle
Trolley jack
Light with extension lead

Buying tools

For practically all tools, a tool factor is the best source since he will have a very comprehensive range compared with the average garage or accessory shop. Having said that, accessory shops often offer excellent quality tools at discount prices, so it pays to shop around.

Remember, you don't have to buy the most expensive items on the shelf, but it is always advisable to steer clear of the very cheap tools. There are plenty of good tools around, at reasonable prices, so ask the proprietor or manager of the shop for advice before making a purchase.

Care and maintenance of tools

Having purchased a reasonable tool kit, it is necessary to keep the tools in a clean and serviceable condition. After use, always wipe off any dirt, grease and metal particles using a clean, dry cloth, before putting the tools away. Never leave them lying around after they have been used. A simple tool rack on the garage or workshop wall, for items such as screwdrivers and pliers is a good idea. Store all normal spanners and sockets in a metal box. Any measuring instruments, gauges, meters, etc., must be carefully stored where they cannot be damaged or become rusty.

Take a little care when the tools are used. Hammer heads inevitably become marked and screwdrivers lose the keen edge on their blades from time-to-time. A little timely attention with emery cloth or a file will soon restore items like this to a good serviceable finish.

Working facilities

Not to be forgotten when discussing tools, is the workshop itself; if anything more than routine maintenance is to be carried out, some form of suitable working area becomes essential.

It is appreciated that many an owner mechanic is forced by circumstance to remove an engine or similar item, without the benefit of a garage or workshop. Having done this, any repairs should always be done under the cover of a roof.

Wherever possible, any dismantling should be done on a clean flat workbench or table at a suitable working height.

Any workbench needs a vice; one with a jaw opening of 4 in (100 mm) is suitable for most jobs. As mentioned previously, some clean dry storage space is also required for tools, as well as the lubricants, cleaning fluids, touch-up paints and so on which soon become necessary.

Another item which may be required, and which has a much more general usage, is an electric drill with a chuck capacity of at least 5/16 in. (8 mm). This, together with a good range of twist drills, is virtually essential for fitting accessories such as wing mirrors and reversing lights.

Last, but not least, always keep a supply of old newspapers and clean, lint-free rags available, and try to keep any working area as clean as possible.

Spanner jaw gap comparison table

Jaw gap (in)	Spanner size
0.250	¼ in AF
0.276	7 mm
0.313	5/16 in AF
0.315	8 mm
0.344	11/32 in AF; 1/8 in Whitworth
0.354	9 mm
0.375	3/8 in AF
0.394	10 mm
0.433	11 mm
0.438	7/16 in AF
0.445	3/16 in Whitworth; ¼ in BSF
0.472	12 mm
0.500	½ in AF
0.512	13 mm
0.525	¼ in Whitworth; 5/16 in BSF
0.551	14 mm
0.562	9/16 in AF
0.591	15 mm
0.600	5/16 in Whitworth; 3/8 in BSF
0.625	5/8 in AF
0.630	16 mm
0.669	17 mm
0.686	11/16 in AF
0.709	18 mm
0.710	3/8 in Whitworth; 7/16 in BSF
0.748	19 mm
0.750	¾ in AF
0.813	13/16 in AF
0.820	7/16 in Whitworth; ½ in BSF
0.866	22 mm
0.875	7/8 in AF
0.920	½ in Whitworth; 9/16 in BSF
0.937	15/16 in AF
0.945	24 mm
1.000	1 in AF
1.010	9/16 in Whitworth; 5/8 in BSF
1.024	26 mm
1.063	1.1/16 in AF; 27 mm
1.100	5/8 in Whitworth; 11/16 in BSF
1.125	1.1/8 in AF
1.181	30 mm
1.200	11/16 in Whitworth; ¾ in BSF
1.250	1¼ in AF
1.260	32 mm
1.300	¾ in Whitworth; 7/8 in BSF
1.313	1.5/16 in AF
1.390	13/16 in Whitworth; 15/16 in BSF
1.417	36 mm
1.438	1.7/16 in AF
1.480	7/8 in Whitworth; 1 in BSF
1.500	1½ in AF
1.575	40 mm; 15/16 in Whitworth
1.614	41 mm
1.625	1.5/8 in AF
1.670	1 in Whitworth; 1.1/8 in BSF
1.688	1.11/16 in AF
1.811	46 mm
1.813	13/16 in AF
1.860	1.1/8 in Whitworth; 1¼ in BSF
1.875	1.7/8 in AF
1.969	50 mm
2.000	2 in AF
2.050	1¼ in Whitworth; 1.3/8 in BSF
2.165	55 mm
2.362	60 mm

Chapter 1 Engine

Contents

Big-end and main bearings - examination and renovation ... 26	General description ... 1
Camshaft and cam-followers - refitting ... 41	Gudgeon pins - removal ... 14
Camshaft and camshaft bearings - examination and renovation ... 29	Lubrication system - description ... 19
Camshaft and tappets - removal ... 11	Main bearings and crankshaft - removal ... 17
Connecting rods - examination and renovation ... 34	Major operations possible with engine in vehicle ... 2
Crankcase ventilation system - description and servicing ... 22	Major operations requiring engine removal ... 3
Crankshaft - examination and renovation ... 25	Method of engine removal ... 4
Crankshaft and rear oilseal - refitting ... 39	Oil filter - removal ... 20
Cylinder bores - examination and renovation ... 27	Oil pump - removal and overhaul ... 21
Cylinder head - decarbonising ... 36	Pistons and connecting rods ... 40
Cylinder head - removal and dismantling ... 8	Pistons and piston rings - examination and renovation ... 28
Cylinder head valves - refitting ... 44	Pistons, connecting rods and big-end bearings - removal ... 13
Distributor - refitting ... 47	Piston rings - removal ... 15
Engine - dismantling general ... 6	Rockers and rocker shaft - examination and renovation ... 32
Engine - initial start up after major overhaul ... 52	Rocker arm/valve - adjustment ... 46
Engine - installation with gearbox ... 51	Rocker shaft and pushrods - refitting ... 45
Engine - installation without gearbox ... 50	Rocker unit - dismantling ... 9
Engine - removal ... 5	Starter ring gear - examination and renovation ... 35
Engine ancillary components - removal ... 7	Sump - removal ... 12
Engine front mountings - removal and installation ... 23	Tappets (cam followers) - examination and renovation ... 33
Engine installation - general ... 49	Timing chain tensioner - removal ... 18
Engine reassembly - general ... 38	Timing chain tensioner and cover - refitting ... 42
Examination and renovation - general ... 24	Timing cover, sprockets and chain - removal ... 10
Fault diagnosis - engine ... 53	Timing sprockets and chain - examination and renovation ... 31
Final assembly ... 48	Valves and valve seats - examination and renovation ... 30
Flywheel - removal ... 16	Valve guides - examination and renovation ... 37
Flywheel/sump/oil pump/crankshaft pulley - refitting ... 43	

Specifications

Engine type ...	'A' (1.3 HC)
Firing order ...	1 - 2 - 4 - 3
Bore ...	3.118 in (80.98 mm)
Stroke ...	2.478 in (62.99 mm)
Swept volume ...	1298 cc
Compression ratio ...	9.0 : 1
Valve actuation ...	Pushrod and rocker arm
Camshaft position ...	In crankcase on right
Pressure @ starter speed ...	142 to 170 lbf/in^2 (10 to 12 kgf/cm^2)
Mean working pressure ...	128 lbf/in^2 (9.0 kgf/cm^2)
Idling speed ...	800 ± 25 rpm
Maximum continuous speed ...	5800 rpm
Engine output (DIN) ...	42 KW (58 PS) (57 HP) @ 5500 rpm
Torque (DIN) ...	91 Nm (9.3 kgf m) (67 lbf ft) @ 3000 rpm

Chapter 1/Engine

Cylinder block

Cast marking on cylinder block:	
1.3 litre	711M-6015-A-A
Number of main bearings	5
Cylinder liner bore	3.311 to 3.314 in (84.112 to 84.175 mm)
Cylinder bore diameter:	
A (standard)	3.1869 to 3.1873 in (80.947 to 80.957 mm)
B	3.1873 to 3.1877 in (80.957 to 80.967 mm)
C	3.1877 to 3.1881 in (80.967 to 80.977 mm)
D	3.1881 to 3.1885 in (80.977 to 80.987 mm)
E	3.1885 to 3.1889 in (80.987 to 80.997 mm)
F	3.1889 to 3.1893 in (80.997 to 81.007 mm)
Bearing width	1.056 to 1.058 in (26.822 to 26.873 mm)
Fitted main bearing shells, vertical internal diameter	2.126 to 2.128 in (54.013 to 54.044 mm)
Undersize - 0.010 in (0.254 mm)	2.116 to 2.118 in (53.759 to 53.790 mm)
Undersize - 0.020 in (0.508 mm)	2.106 to 2.108 in (53.505 to 53.536 mm)
Undersize - 0.030 in (0.762 mm)	2.096 to 2.098 in (53.251 to 53.282 mm)
Main bearing bore (in block):	
Standard	2.271 to 2.2715 in (57.683 to 57.696 mm)
Oversize	2.286 to 2.2865 in (58.064 to 58.077 mm)
Camshaft bearing bore (in block):	
Standard	1.6885 to 1.6894 in (42.888 to 42.913 mm)
Oversize	+ 0.020 in (+ 0.508 mm)

Crankshaft

Main bearing journal diameter:	
Standard	2.125 to 2.126 in (53.983 to 54.003 mm)
Undersize - 0.010 in (0.254 mm)	2.115 to 2.116 in (53.729 to 53.749 mm)
Undersize - 0.020 in (0.508 mm)	2.105 to 2.106 in (53.475 to 53.495 mm)
Undersize - 0.030 in (0.762 mm)	2.095 to 2.096 in (53.221 to 53.241 mm)
Crankshaft endfloat	0.003 to 0.011 in (0.075 to 0.280 mm)
Length of main bearing shell	0.995 to 1.005 in (25.273 to 25.527 mm)
Play in main journal bearing shell	0.0004 to 0.0024 in (0.010 to 0.061 mm)
Crank pin diameter - Standard	1.937 to 1.938 in (49.195 to 49.215 mm)
Undersize - 0.002 in (0.05 mm)	1.935 to 1.936 in (49.144 to 49.164 mm)
Undersize - 0.010 in (0.25 mm)	1.927 to 1.928 in (48.941 to 48.961 mm)
Undersize - 0.020 in (0.51 mm)	1.917 to 1.918 in (48.687 to 48.707 mm)
Undersize - 0.030 in (0.76 mm)	1.907 to 1.908 in (48.433 to 48.453 mm)
Undersize - 0.040 in (1.02 mm)	1.897 to 1.898 in (48.179 to 48.199 mm)

Camshaft

Marking	711 F-6250 cc
Thickness of camshaft retaining plate	0.1760 in (4.470 mm)
Cam lift:	
Inlet	0.236 in (5.985 mm)
Exhaust	0.232 in (5.894 mm)
Length of cams (between heel and tip):	
Inlet	1.303 in (33.087 mm)
Exhaust	1.312 in (33.326 mm)
Drive	By chain with tensioning device
Camshaft bearing diameter - Front, centre, rear	1.560 to 1.561 in (39.616 to 39.637 mm)
Internal diameter of bearing bush - Front, centre, rear	1.561 to 1.562 in (39.662 to 39.675 mm)
Camshaft endfloat	0.002 to 0.008 in (0.06 to 0.2 mm)

Pistons

Piston diameter:	
Grade E - Standard	3.1872 to 3.1876 in (80.954 to 80.964 mm)
Grade F	3.1876 to 3.1880 in (80.964 to 80.974 mm)
Piston diameter (Oversize)	0.003 in (0.064 mm)
Grade E	3.1890 to 3.1901 in (81.018 to 81.028 mm)
Grade F	3.1901 to 3.1904 in (81.028 to 81.038 mm)
Piston to bore clearance	0.0009 to 0.0010 in (0.023 to 0.043 mm)
Ring gap (fitted in block):	
Top	0.009 to 0.014 in (0.23 to 0.36 mm)
Centre	0.009 to 0.014 in (0.23 to 0.36 mm)
Bottom	0.009 to 0.014 in (0.23 to 0.36 mm)

Gudgeon pins

Length of gudgeon pin	2.795 to 2.810 in (70.99 to 71.37 mm)
Pin diameter:	
1	0.8119 to 0.8120 in (20.622 to 20.625 mm)
2	0.8120 to 0.8121 in (20.625 to 20.627 mm)

Chapter 1/Engine

3	0.8121 to 0.8122 in (20.627 to 20.630 mm)
4	0.8122 to 0.8123 in (20.630 to 20.632 mm)
Pin interference in piston at 21°C (70°F)	0.0001 to 0.0003 in (0.003 to 0.008 mm)
Clearance in connecting rod at 21°C (70°F)	0.00015 to 0.0004 in (0.004 to 0.010 mm)

Connecting rods

Bore diameter of big-end	2.0823 to 2.0831 in (52.89 to 52.91 mm)
Bore diameter of small end:	
White	0.8122 to 0.8123 in (20.629 to 20.632 mm)
Red	0.8123 to 0.8124 in (20.632 to 20.634 mm)
Yellow	0.8124 to 0.8125 in (20.634 to 20.637 mm)
Blue	0.8125 to 0.1826 in (20.637 to 20.640 mm)
Vertical internal diameter:	
Standard	1.938 to 1.939 in (49.221 to 49.260 mm)
Undersize 0.002 in (0.051 mm)	1.936 to 1.937 in (49.170 to 49.208 mm)
Undersize - 0.010 in (0.254 mm)	1.928 to 1.929 in (48.967 to 49.005 mm)
Undersize - 0.020 in (0.508 mm)	1.918 to 1.919 in (48.713 to 48.751 mm)
Undersize - 0.030 in (0.762 mm)	1.909 to 1.913 in (48.491 to 48.592 mm)
Undersize - 0.040 in (1.016 mm)	1.898 to 1.899 in (48.205 to 48.243 mm)
Clearance - big-end journal to bearing	0.0002 to 0.003 in (0.006 to 0.064 mm)

Cylinder head

Cast marking on cylinder head:	
1.3 litre	A
Valve seat angle in head	46°
Stem bore, inlet and exhaust valves	0.311 to 0.312 in (7.907 to 7.937 mm)
Bore for bushes	0.438 to 0.439 in (11.133 to 11.153 mm)

Valves

Valve clearances (cold):	
Inlet (Engine code J3)	0.010 in (0.25 mm)
Inlet (other engine codes)	0.008 in (0.20 mm)
Exhaust	0.022 in (0.56 mm)
Inlet valve:	
Opens	21° BTDC
Closes	55° ABDC
Exhaust valve:	
Opens	70° BBDC
Closes	22° ATDC
Valve springs (number of turns)	3.75 to 5.75
Cam follower diameter	0.515 to 0.516 in (13.081 to 13.094 mm)
Clearance (cam follower to block)	0.0005 to 0.0019 in (0.013 to 0.05 mm)

Inlet valves

Length	4.357 to 4.396 in (110.67 to 111.67 mm)
Valve head diameter:	
1.3 litre	1.497 to 1.507 in (38.02 to 38.28 mm)
Valve stem diameter:	
Standard	0.3097 to 0.3104 in (7.868 to 7.886 mm)
Oversize - 0.003 in (0.076 mm)	0.3128 to 0.3135 in (7.945 to 7.962 mm)
Oversize - 0.015 in (0.381 mm)	0.3248 to 0.3255 in (8.249 to 8.267 mm)
Valve stem play in guide	0.0008 to 0.0027 in (0.02 to 0.068 mm)
Valve lift	0.33 in (8.38 mm)

Exhaust valves

Length	4.345 to 4.365 in (110.36 to 110.87 mm)
Valve cup diameter:	
1.3 litre.	1.234 to 1.244 in (31.34 to 31.59 mm)
Valve stem diameter:	
Standard	0.3089 to 0.3096 in (7.846 to 7.863 mm)
Oversize - 0.003 in (0.076 mm)	0.3119 to 0.3126 in (7.922 to 7.939 mm)
Oversize - 0.015 in (0.381 mm)	0.3239 to 0.3245 in (8.227 to 8.243 mm)
Valve stem play in guide	0.0017 to 0.0036 in (0.043 to 0.091 mm)
Valve lift	0.325 in (8.25 mm)

Engine lubrication

Oil type	HD oil
Viscosity:	
under −12°C	SAE 5W/20
under 0°C	SAE 5W/30
−23°C to +32°C	SAE 10W/30, SAE10W/40 or SAE 10W/50
over −12°C	SAE 20W/40 or SAE 20W/50
Ford specification	SS-M2C-9001AA

Initial capacity with filter	6.5 pints (3.67 litres)
Oil change without filter change	4.8 pints (2.75 litres)
Oil change with filter change	5.7 pints (3.25 litres)
Minimum oil pressure at:	
700 rpm and 80°C	8.5 lbf/in^2 (0.6 kp/cm^2)
2000 rpm and 80°C	21 lbf/in^2 (1.5 kp/cm^2)
Oil pressure warning light glows at	6 ± 1.5 lb f/in^2 (0.4 ± 0.1 kp/cm^2)
Excess pressure valve opens at	35 to 40 lb f/in^2 (2.46 to 2.81 kp/cm^2)
Oil pump play with external rotor casing	0.0055 to 0.0105 in (0.1397 to 0.2667 mm)
Gap internal/external rotor	0.002 to 0.005 in (0.0508 to 0.1270 mm)
Axial play of external and internal rotor in relation to oil pump cover	0.001 to 0.0025 in (0.0254 to 0.0635 mm)

Torque wrench settings

	lb f ft	kg f m
Main bearing caps	49 to 55	6.8 to 7.6
Connecting rod bolts	38 to 43	5.3 to 5.9
Crankshaft belt pulley	40 to 45	5.5 to 6.2
Camshaft chain sprocket	40 to 44	5.5 to 6.1
Rear sealing ring carrier	13 to 15	1.7 to 2.1
Flywheel	50 to 56	6.8 to 7.6
Clutch thrust plate to flywheel	13 to 15	1.7 to 2.1
Front crankcase cover	15 to 18	1.5 to 1.8
Oil pump	13 to 15	1.7 to 2.1
Oil pump inlet pipe	13 to 15	1.7 to 2.1
Oil pump cover	6 to 9	0.8 to 1.2
Rocker shaft	18 to 22	2.4 to 3.0
Rocker cover	2 to 3.6	0.3 to 0.5
Cylinder head:		
(1)	7	1.0
(2)	22 to 36	3.0 to 5.0
(3)	59 to 66	8.2 to 9.2
(4) after 10 to 20 minutes wait	78 to 85	11.0 to 11.7
(5) after engine has warmed up (15 minutes at 1000 rpm) tighten up	78 to 85	11.0 to 11.7
Sump:		
(1)	3 to 5	0.4 to 0.7
(2)	6 to 8	0.8 to 1.1
Oil drain screw	20 to 25	2.7 to 3.4
Oil pressure switch	10 to 11	1.3 to 1.5
Spark plugs	22 to 29	3.0 to 3.9
Inlet manifold	13 to 15	1.7 to 2.1
Exhaust manifold	15 to 18	2.1 to 2.5
Fuel pump	12 to 15	1.63 to 2.03
Water pump	5 to 7	0.7 to 1.0
Thermostat housing	13 to 15	1.7 to 2.1
Fan to water pump flange	5 to 7	0.7 to 1.0
Timing chain tensioner	5 to 7	0.7 to 1.0

Fig. 1.1. The 1300cc Capri II engine unit

1 General description

The power unit is of the four cylinder in-line overhead valve type engine. The cross flow cylinder head is of cast iron construction and the vertically mounted valves run in guides cast integrally with the cylinder head.

The valves are operated by rocker arms, pushrods and tappets (cam followers) from lobes on the camshaft which is located within the crankcase on the right-hand side.

Valve clearance is obtained by means of self-locking adjusting screws.

The crankcase and cylinder block are also of cast iron construction and the bottom of the crankcase is enclosed by a pressed steel sump.

Light alloy pistons are used, having two compression and one oil control ring.

The gudgeon pin is retained in position by circlips engaged in grooves in the piston. At the front of the engine, a single chain drives the camshaft through sprockets all of which are enclosed in a pressed steel timing cover.

The tension of the chain is maintained automatically by a small cam which bears against the tensioner arm.

The camshaft runs in three renewable bearings and camshaft endfloat is controlled by a thrust plate.

The cast iron crankshaft is carefully balanced during production and is supported in five shell type main bearings. Crankshaft endfloat is controlled by semi-circular thrust washers which are located on either side of the centre main bearing.

A pulley on the front end of the crankshaft drives the water pump, fan and alternator through a vee drivebelt.

The distributor is driven from a skew gear on the camshaft.

The externally mounted oil pump is also driven from the same camshaft skew gear as the distributor.

The flywheel is mounted on the crankshaft rear flange and to the flywheel is bolted the clutch mechanism.

2 Major operations possible with engine in vehicle

The following major operations can be carried out to the engine with it in place in the bodyframe. Removal and replacement of the:

1. Cylinder head assembly
2. Oil pump
3. Engine front mountings
4. Engine/gearbox rear mounting

3 Major operations requiring engine removal

The following major operations can be carried out with the engine out of the bodyframe and on the bench or floor. Removal and replacement of the:

1. Sump
2. Big-end bearings
3. Pistons and connecting rods
4. Main bearings
5. Crankshaft
6. Flywheel
7. Crankshaft rear bearing oil seal
8. Camshaft
9. Timing chain and gears

4 Method of engine removal

1. The easiest way to remove the engine is to remove it on its own leaving the gearbox in position in the car.
2. If heavy duty lifting tackle is available, then the engine/gearbox can be removed together in which case the gearbox attachments will have to be disconnected as described in Chapter 6.

5 Engine - removal

1. Obtain a suitable hoist and a jack.
2. Open the bonnet to its fullest extent and with the help of an assistant, unbolt the hinges and lift it from the car.
3. Disconnect the lead from the battery negative terminal, and then remove the air cleaner unit from the carburettor (photo).
4. Disconnect the engine earth strap.
5. Remove the under sump shield.
6. Drain the coolant by disconnecting the bottom hose from the radiator (no drain tap being fitted) and also opening the cylinder block drain plug which is located on the left-hand side toward the rear.

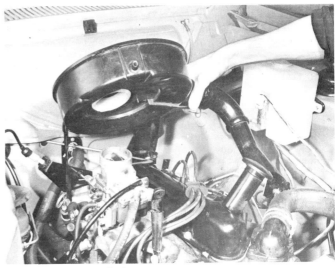

5.3 Lifting away the air cleaner

Fig. 1.2. The sump shield retaining bolts (arrowed) (Sec. 5)

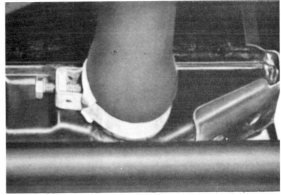

Fig. 1.3. The bottom hose location on the radiator (Sec. 5)

Retain the coolant if the anti-freeze mixture is still suitable for re-use.
7 Disconnect the radiator top hose (photo) and the heater hoses.
8 Remove the air deflector panel and fan shroud.
9 Unbolt and remove the radiator from the engine compartment.
10 Disconnect the water temperature sender unit (photo), and also remove the fan and pulley (photo).
11 Disconnect the lead from the oil pressure switch.
12 Disconnect the leads from the rear of the alternator.
13 Disconnect the leads from the starter motor (photo).
14 Disconnect the HT and LT leads which run between the coil and the distributor.
15 Disconnect the throttle cable from the carburettor and unbolt its support bracket from the inlet manifold (photo), and unclip the throttle shaft from the carburettor (photo).
16 Disconnect the choke cable from the carburettor (photo).
17 Disconnect the fuel inlet pipe from the fuel pump.
18 Unscrew the three mounting bolts and remove the starter motor.
19 Disconnect the exhaust downpipe from the exhaust manifold.
20 From the lower front face of the clutch bellhousing remove the cover plate.
21 Remove the bolts which secure the clutch bellhousing to the engine crankcase.
22 Attach a hoist to the engine and just take its weight, then disconnect the engine front mountings.
23 Place a jack under the gearbox making sure that it is located securely to support the gearbox as the engine is withdrawn.
24 Raise the hoist slightly and pull the engine forward to disconnect it from the gearbox.
25 Lift the engine upwards and out of the engine compartment.

6 Engine - dismantling (general)

1 It is best to mount the engine on a dismantling stand but if one is not available, then stand the engine on a strong bench at a comfortable working height.
2 During the dismantling process the greatest care should be taken to keep the exposed parts free from dirt. As an aid to achieving this, it is sound advice to thoroughly clean down the outside of the engine, removing all traces of oil and congealed dirt.
3 Use paraffin or a good grease solvent. The latter compound will make the job much easier, as, after the solvent has been applied and allowed to stand for a time, a vigorous jet of water will wash off the solvent and all the grease and filth. If the dirt is thick and deeply embedded, work the solvent into it with a stiff paintbrush.
4 Finally, wipe down the exterior of the engine with a rag and only then, when it is quite clean, should the dismantling process begin. As the engine is stripped, clean each part in a bath of paraffin or petrol.
5 Never immerse parts with oilways in paraffin, eg. the crankshaft, but to clean, wipe down carefully with a petrol dampened rag. Oilways can be cleaned out with wire. If an air line is present all parts can be blown dry and the oilways blown through as an added precaution.
6 Re-use of old engine gaskets is false economy, and can give rise to oil and water leaks, if nothing worse. To avoid the possibility of trouble after the engine has been reassembled **always** use new gaskets throughout.
7 Do not throw away the old gaskets as it sometimes happens that an immediate replacement cannot be found and the old gasket is then very useful as a template. Hang up the old gaskets as they are removed on a suitable hook or nail.
8 To strip the engine it is best to work from the top down. The sump provides a firm base on which the engine can be supported in an upright position. When the stage where the sump must be removed is reached, the engine can be turned on its side and all other work carried out with it in this position.
9 Wherever possible, replace nuts, bolts and washers fingertight from wherever they were removed. This helps avoid later loss and muddle. If they cannot be replaced then lay them out in such a fashion that it is clear from where they came.

7 Engine ancillary components - removal

1 Before basic engine dismantling begins the engine should be stripped of all its ancillary components. These items should also be removed if a factory exchange reconditioned unit is being purchased.

The items comprise:

Alternator and brackets
Water pump and thermostat housing
Distributor and spark plugs
Inlet and exhaust manifold and carburettor
Fuel pump and fuel pipes
Oil filter and dipstick
Oil filler cap
Clutch assembly (Chapter 5)
Engine mountings
Oil pressure sender unit
Water temperature sender unit

2 Without exception all these items can be removed with the engine in the car if it is merely an individual item which requires attention. (It is necessary to remove the gearbox if the clutch is to be renewed with the engine in position).
3 Remove the alternator after undoing the nuts and bolts which secure it in place. Remove the alternator mounting bracket (photo).
4 Remove the distributor by disconnecting the vacuum pipe, unscrew the single bolt at the clamp plate and lift out the distributor (photo).
5 Remove the oil pump assembly by unscrewing the three securing bolts with their lockwashers (photo).
6 Unscrew the two bolts securing the fuel pump (photo).
7 Unscrew the oil pressure sender unit (photo).
8 Remove the inlet and exhaust manifolds together with the carburettor by undoing the bolts and nuts which hold the units in place.
9 Unbolt the securing bolts of the water elbow and lift out the thermostat (photos).
10 Bend back the tab lockwashers, where fitted, and undo the bolts which hold the water pump and engine mountings in place.
11 Undo the bolts holding the clutch cover flange to the flywheel a third of a turn each in a diagonal sequence, repeating this operation until the clutch and driven plate can be lifted off (photo).
12 Loosen the clamp securing the rubber tube from the oil separator unit to the inlet manifold and pull off the tube. Remove the oil separator location on the fuel pump mounting pad by carefully prising it off, after first unscrewing the bolt retaining the separator to the block.
13 The engine is now stripped of ancillary components and ready for major dismantling to begin.

8 Cylinder head - removal and dismantling

1 If the engine is still in the car, drain the cooling system, disconnect the top radiator hose and the heater hoses. Disconnect the battery and all leads and controls from the cylinder head. Disconnect the exhaust downpipe from the manifold.
2 Remove the four screws and flat washers which hold the flange of the rocker cover to the cylinder head and lift off the rocker cover and gasket (photo).
3 Unscrew the four rocker shaft pedestal bolts evenly and remove together with their washers.
4 Lift off the rocker assembly as one unit (photo).
5 Remove the pushrods, keeping them in the relative order in which they were removed. The easiest way to do this is to push them through a sheet of thick paper or thin card in the correct sequence (photo).
6 Undo the cylinder head bolts half a turn at a time in the reverse order to that shown in Fig. 1.25. When all the bolts are no longer under tension they may be unscrewed from the cylinder head one at a time (photo).
7 The cylinder head can now be removed by lifting upward. If the head is jammed, try to rock it to break the seal. Under no circumstances try to prise it apart from the block with a screwdriver or cold chisel as damage may result to the faces of the head or block. If the head will not readily free, turn the engine over by applying a spanner to the pulley bolt and the compression in the cylinders will often break the cylinder head joint. If this fails to work, strike the head sharply with a plastic headed hammer, or with a wooden hammer, or with a metal hammer with an interposed piece of wood to cushion the blows. Under no circumstances hit the head directly with a metal hammer as this may cause the iron casting to fracture. Several sharp taps with the hammer at the same time pulling upward should free the head (photo).

5.7 Disconnect the radiator top hose

5.10a Upper left: water temperature removal; Centre: engine earth strap

5.10b Remove the fan and fan pulley

5.13 Unscrew the nut and remove the cable from the starter motor

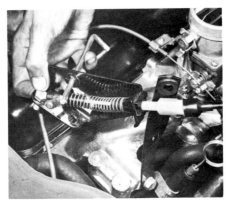
5.15a Detach the throttle cable bracket

5.15b Unclip the throttle shaft from the carburettor

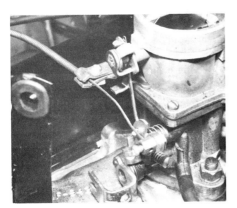
5.16 Detach the choke cable - inner and outer from the carburettor

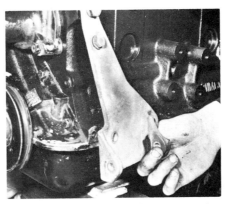
7.3 Removing the alternator mounting bracket

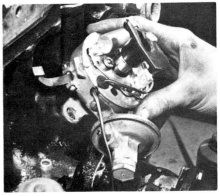
7.4 Remove the distributor

7.5 Oil pump retaining bolts (arrowed). Note: the oil filter has been removed

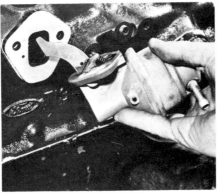
7.6 Remove the fuel pump from the cylinder block

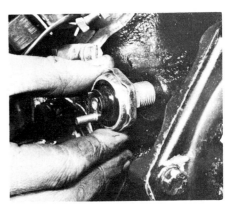
7.7 Unscrew the oil pressure sender unit from the cylinder block after removing the sender unit lead

Chapter 1/Engine

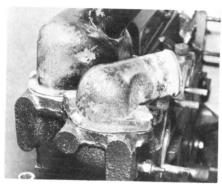

7.9a Removing the water elbow to expose the thermostat ...

7.9b ... which can then be lifted from its locations in the cylinder head

7.11 Lifting the clutch pressure plate and disc off of the flywheel after first removing the pressure plate bolts. Note dowel pips in flywheel to ensure correct location of plate

8.2 Removing the rocker cover to expose the valve train. Note the dovetails on the rocker cover and gasket to assist in gasket location

8.4 Lifting off the rocker shaft assembly from the cylinder head

8.5 Removing the valve pushrods. Pushrods must be refitted in the order that they were removed from the cylinder head

8.6 Removing the cylinder head bolts. All bolts must be slackened off before any bolt is removed

8.7 Lifting the cylinder head off the cylinder block

8.9 A valve spring partially compressed showing collets (arrowed) ready for removal

8 Do not lay the cylinder head face downward unless the plugs have been removed as they protrude and can be easily damaged.
9 The valves can be removed from the cylinder head by compressing each spring in turn with a valve spring compressor until the two halves of the collets can be removed. Release the compressor and remove the spring and spring retainer (photo).
10 If, when the valve spring compressor is screwed down, the valve spring retaining cap refuses to free to expose the split collet, do not continue to screw down on the compressor as there is a likelihood of damaging the valve.
11 Gently tap the top of the tool directly over the cap with a light hammer. This will free the cap. To avoid the compressor jumping off the valve spring retaining cap when it is tapped, hold the compressor firmly in position with one hand.
12 Slide the rubber oil control seal off the top of each inlet valve stem and then drop out each valve through the combustion chamber (photo).
13 It is essential that the valves are kept in their correct sequence unless they are so badly worn that they are to be renewed. If they are going to be kept and used again, place them in a sheet of card having eight holes numbered 1 to 8 corresponding with the relative positions the valves were in when originally installed. Also keep the valve springs, washers and collets in their original sequence.

9 Rocker shaft - dismantling

1 Pull out the split pin from each end of the rocker shaft and remove the flat washer, crimped spring washer and the remaining flat washer.
2 The rocker arms, rocker pedestals, and distance springs can now be slid off the end of the shaft.
3 If the original parts are to be reassembled, keep them in the sequence in which they were removed.

10 Timing cover, sprockets and chain - removal

1 The timing cover, sprockets and chain can be removed with the engine in the car, provided the radiator, fan and water pump are first

Fig. 1.4. Cylinder head, crankcase and sump components

1 Cylinder head
2 Distributor
3 Oil separator (ventilation system)
4 Fuel pump
5 Oil pressure switch
6 Spigot bearing
7 Rear oil seal carrier
8 Oil pump
9 Oil filter
10 Pick up pipe
11 Water outlet elbow
12 Thermostat
13 Water pump
14 Camshaft
15 Thrust plate
16 Camshaft timing gear
17 Timing chain
18 Front cover and oil seal
19 Timing chain tensioner
20 Crankshaft

8.12 Valve stem seal: seal is removed by sliding up valve stem. Note collet retaining grooves at top of valve stem

10.2a The crankshaft pulley centre bolt (arrowed) ...

10.2b ... must be removed before the pulley can be detached

Fig. 1.5. Removing the crankshaft pulley using two leg puller (Sec. 10)

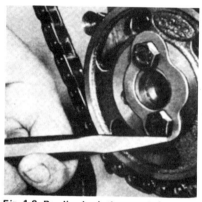

Fig. 1.6. Bending back the camshaft timing gear retaining nut locktabs (Sec. 10)

10.8 Easing the crankshaft and camshaft sprockets from their respective stubs complete with timing chain

12.2 Remove the sump bolts and ...

12.3 ... detach the sump from underside of block

13.2 Big-end bearing cap retaining bolts (arrowed)

removed (see Chapter 2).

2 Unscrew the bolt from the centre of the crankshaft pulley (photos). The best way to do this is to fit a ring spanner and then give it a sharp blow with a soft-faced hammer in an anticlockwise direction. Alternatively, engage a gear and apply the handbrake fully to prevent the engine turning when the spanner is turned.

3 The crankshaft pulley wheel may pull off quite easily. If not, place two large screwdrivers behind the wheel at 180º to each other, and carefully lever off the wheel. It is preferable to use a proper pulley extractor if this is available, but large screwdrivers or tyre levers are quite suitable, providing care is taken not to damage the pulley flange.

4 Undo the bolts which hold the timing cover in place, noting that four front sump bolts must also be removed before the cover can be taken off.

5 Check the chain for wear by measuring how much it can be depressed. More than ½ inch (12.5 mm) means a new chain must be fitted on reassembly.

6 With the timing cover off, take off the oil thrower. Note that the concave side faces outward.

7 With a drift or screwdriver tap back the tabs on the lockwasher under the two camshaft sprocket retaining bolts and undo the bolts.

8 To remove the camshaft and crankshaft timing wheels complete with chain, ease each wheel forward a little at a time, levering behind each sprocket wheel in turn with two large screwdrivers at 180º to each other. If the gearwheels are locked solid then it will be necessary to use a proper pulley extractor, and if one is available this should be used in preference to screwdrivers. With both sprocket wheels safely off, remove the Woodruff key from the crankshaft with a pair of pliers (photo).

11 Camshaft and tappets - removal

1 The camshaft can only be removed from the engine when the engine is removed from the vehicle. This is due to the fact that in order to remove and refit the tappets, the engine must be inverted.

2 With the engine inverted and sump, rocker gear, pushrods, timing

Chapter 1/Engine

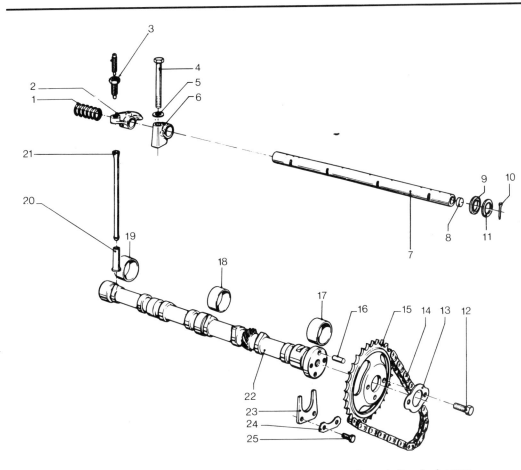

Fig. 1.7 Component parts of camshaft and valve gear

1 Rocker arm spring
2 Rocker arm
3 Adjusting screw
4 Bolt
5 Washer
6 Pedestals
7 Rocker shaft
8 Plug
9 Spacer
10 Split pin
11 Shim
12 Bolt
13 Tab washer
14 Timing chain
15 Sprocket
16 Dowel pin
17 Bearing
18 Bearing
19 Bearing
20 Cam follower
21 Push rod
22 Camshaft
23 Thrust plate
24 Tab washer
25 Bolt

cover, oil pump, gearwheels and timing chain removed, take off the chain tensioner and arm.

3 Knock back the lockwasher tabs from the two bolts which hold the 'U' shaped camshaft retainer in place behind the camshaft flange and slide out the retainer.

4 Rotate the camshaft so that the tappets are fully home and then withdraw the camshaft from the cylinder block. Take great care that the cam lobe peaks do not damage the camshaft bearings as the shaft is pulled forward.

5 Remove the tappets (cam followers) and keep them in their original fitted sequence.

12 Sump - removal

1 Although it is not impossible to remove the sump with the engine in the car, it is considered completely impracticable due to the necessity to detach the steering and suspension, and remove the front crossmember. The procedure described assumes that the engine has been removed and the gearbox and starter motor detached.
2 Unscrew and remove the sump securing bolts (photo).
3 Remove the sump and then clean away all pieces of gasket from the crankcase and sump flanges (photo).

13 Pistons, connecting rods and big-end bearings - removal

1 The pistons and connecting rods can be removed with the engine on the bench and lying on its side.
2 With the cylinder head and sump removed, undo the big-end retaining bolts (photo).
3 The connecting rods and pistons are lifted out through the top of the cylinder block.

Fig. 1.8. Method of numbering connecting rods (arrowed) and main bearing caps (Sec. 13)

4 Remove the big-end caps, one at a time, noting that they are numbered 1 to 4 with matched cap numbers, so that exact refitting will be facilitated (Fig. 1.8) (photo).
5 Keep the original shell bearings with each connecting rod. Should the big-end caps be difficult to remove, then they may be tapped gently using a soft-faced mallet.
6 The shell bearings may be removed from the big-end caps and the connecting rods by inserting a thin screwdriver at the shell locating notch.
7 As each piston/connecting rod assembly is withdrawn, mark it so that it will be returned to its original bore. Temporarily refit the big-end caps to the connecting rods to reduce the risk of mixing them up.

13.4 Detaching the big-end cap from rod and crankshaft journal

14.1 Removing the gudgeon pin circlip

16.3 Flywheel bolt location

17.2a Removing a main bearing cap

17.2b The main bearing caps. Note markings ie. 'F', 'R2', 'C', 'R4' and 'R'

17.3 Removing the main bearing thrust washers from their locations on each side of the centre main bearing

18.1 Removing the two bolts retaining the timing chain tensioner to the underside of the block

18.2 Timing chain tensioner arm on tensioner hinge pin

14 Gudgeon pins - removal

1 To remove the gudgeon pin to free the piston from the connecting rod, remove one of the circlips at either end of the pin with a pair of circlip pliers (photo).
2 Press out the pin from the rod and piston.
3 If the pin shows reluctance to move, then on no account force it out, as this could damage the piston. Immerse the piston in a pan of boiling water for three minutes. On removal the expansion of the aluminium should allow the gudgeon pin to slide out easily.
4 Ensure that each gudgeon pin is kept with the piston from which it was removed for exact refitting.

15 Piston rings - removal

1 To remove the piston rings, slide them carefully over the top of the piston, taking care not to scratch the aluminium alloy. Never slide them off the bottom of the piston skirt. It is very easy to break the iron piston rings if they are pulled off roughly so this operation should be done with extreme caution. It is useful to employ three strips of thin metal or feeler gauges to act as guides to assist the rings to pass over the empty grooves and to prevent them from dropping in.
2 Lift one end of the piston ring to be removed out of its groove and insert the end of the feeler gauge under it.
3 Turn the feeler gauges slowly round the piston and as the ring comes out of its groove apply a light upward pressure so that it rests on the land above. It can then be eased off the piston.

16 Flywheel - removal

1 Remove the clutch (Chapter 5).
2 No lock tabs are fitted under the six bolts which hold the flywheel to the flywheel flange on the rear of the crankshaft.
3 Unscrew the bolts and remove them (photo).

4 Lift the flywheel away from the crankshaft flange.
Note: *Some difficulty may be experienced in removing the bolts through the rotation of the crankshaft every time pressure is put on the spanner. To lock the crankshaft in position while the bolts are removed, wedge a block of wood between the crankshaft web and the inside of the crankcase.*

17 Main bearings and crankshaft - removal

1 Unscrew each of the ten bolts securing the five crankshaft main bearing caps and remove them.
2 Remove each main bearing cap in turn noting that they are marked so that there can be no confusion when refitting regarding sequence or orientation (photos). Arrows point towards front of engine.
3 Remove the semi-circular thrust washers fitted each side of the centre main bearing (photo).
4 Lift the crankshaft from the crankcase and then withdraw the shell bearing halves from the crankcase recesses.

18 Timing chain tensioner - removal

1 Undo the two bolts and washers which hold the timing chain tensioner in place. Lift off the tensioner (photo).
2 Pull the timing chain tensioner arm off its hinge pin on the front of the block (photo).

19 Lubrication system - description

1 A forced feed system of lubrication is used with oil circulated round the engine by a pump drawing oil from the sump below the block (Fig. 1.10).
2 The full flow filter and oil pump assembly is mounted externally on the right-hand side of the cylinder block. The pump is driven by means of a short shaft and skew gear from the camshaft.
3 Oil reaches the pump via a tube pressed into the cylinder block sump face. Initial filtration is provided by a spring loaded gauze on the

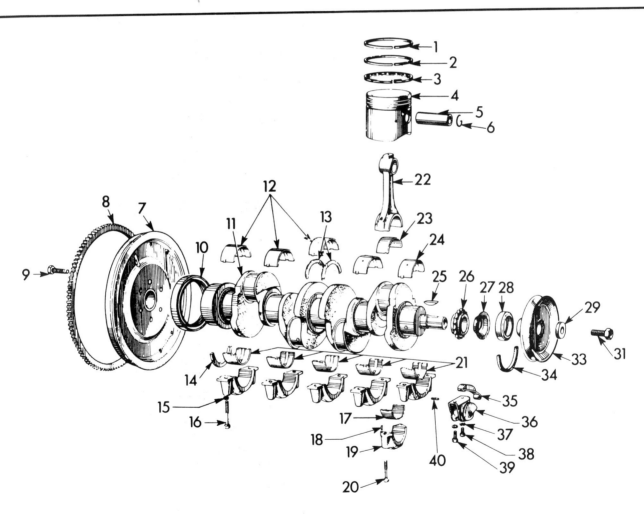

Fig. 1.9. The crankshaft and piston component parts

1 Piston ring (compression)	11 Crankshaft	21 Main bearing shells	31 Bolt
2 Piston ring (compression)	12 Main bearing shells	22 Connecting rod	33 Crankshaft pulley
3 Piston ring (oil control)	13 Thrust washers	23 Big-end shell	34 Seal
4 Piston	14 Seal	24 Main bearing shell	35 Timing chain tensioner
5 Gudgeon pin	15 Spring washer	25 Woodruff key	36 Tensioner ratchet assembly
6 Circlip	16 Set screw	26 Timing chain	37 Spring washer
7 Flywheel	17 Big-end bearing shell	27 Oil thrower	38 Screw
8 Starter - ring gear	18 Dowel	28 Oil seal	39 Screw
9 Bolt	19 Big-end bearing cap	29 Spacer	40 Swivel pin
10 Oil seal	20 Set screw		

Chapter 1/Engine

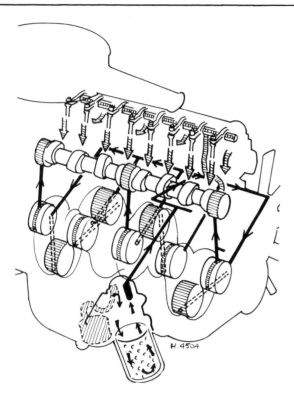

Fig. 1.10. The engine lubrication circuit (Sec. 19).

Fig. 1.11. Removing the oil filter using a wrench (Sec. 20)

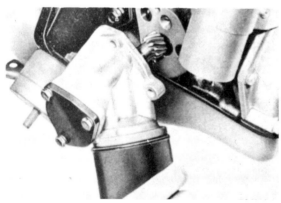

Fig. 1.12. Remove the oil pump (Sec. 21)

end of the tube. Drillings in the block carry the oil under pressure to the main and big-end bearings. Oil at a reduced pressure is fed to the valve and rocker gear and the timing chain and gearwheels.

20 Oil filter - removal and refitting

1 A full-flow type oil filter is located adjacent to the oil pump on the right-hand side of the engine block.
2 This is a cartridge type filter which screws directly into the underside of the pump assembly.
3 Before unscrewing the filter from the pump remember to position a drain tray to catch any oil spillage.
4 An oil filter wrench will probably be required to unscrew the old unit, but if one is not available, drive a screwdriver through the oil filter casing (near its bottom end) and use this as a lever to unscrew it.
5 Smear the sealant ring of the replacement filter with clean oil, and then screw the filter to the pump until handtight.

21 Oil pump - removal and overhaul

1 If the pump is worn it it best to purchase an exchange reconditioned unit as a good oil pump is the very heart of a long engine life. Generally speaking, an exchange or overhauled pump should be fitted at a major engine reconditioning. If it is preferred to overhaul the oil pump, detach the pump and filter unit from the cylinder block, and remove the filter cartridge.
2 Unscrew and remove the four bolts and lockwashers which secure the oil pump cover and remove the cover. Lift out the 'O' ring seal from the groove in the pump body.
3 Check the clearance between the inner and outer rotors with a feeler gauge. This should not exceed 0.006 inch (0.15 mm).
4 Check the clearance between the outer rotor and the pump body. This should not exceed 0.010 inch (0.25 mm).
5 Check the endfloat of the pump rotors by placing a straightedge across the open face of the pump casing and measuring the gap between its lower edge and the face of the rotor. This should not exceed 0.005 inch (0.1270 mm).
6 Replacement rotors are only supplied as a matched pair so, if the clearance is excessive, a new rotor assembly must be fitted. When it is necessary to renew the rotors, drive out the pin securing the skew gear and pull the gear from the shaft. Remove the inner rotor and drive shaft and withdraw the outer rotor. Install the outer rotor with the chamfered end towards the pump body.
7 Fit the inner rotor and driveshaft assembly, position the skew gear and install the pin. Tap over each end of the pin to prevent it loosening in service. Position a new 'O' ring in the groove in the pump body, fit the end plate in position and secure with the four bolts and lockwashers.

22 Crankcase ventilation system - description and servicing

1 A semi-closed positive ventilation system is fitted. A breather valve in the oil filler cap allows air to enter as required. Crankcase fumes travel out through an oil separator and emission control valve, and then via a connecting tube back into the inlet manifold. In this way the majority of crankcase fumes are burnt during the combustion process in the cylinder.
2 With this emission control type system, clean the valve and rocker box cover breather cap every 18,000 miles (29,000 km). To remove the valve, disconnect the hose and then pull it from its grommet in the oil separator box.
3 Dismantle the valve by removing the circlip and extracting the seal, valve and spring from the valve body. (Fig. 1.18).
4 Wash and clean all components in petrol to remove sludge or deposits, and renew the rubber components if they have deteriorated.
5 Reassembly and refitting are reversals of removal and dismantling procedures.

23 Engine front mountings - removal and installation

1 With time the bonded rubber insulators, one on each of the front mountings, will perish causing undue vibration and noise from the

Chapter 1/Engine

Fig. 1.13. Measure clearance between inner and outer oil pump rotors (Sec. 21)

Fig. 1.14. Measure clearance between outer rotor and oil pump body (Sec. 21)

Fig. 1.15. Measure oil pump endfloat (Sec. 21)

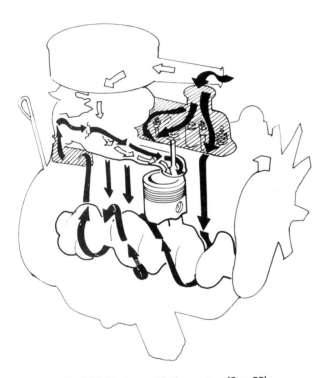

Fig. 1.16. Engine ventilation system (Sec. 22)

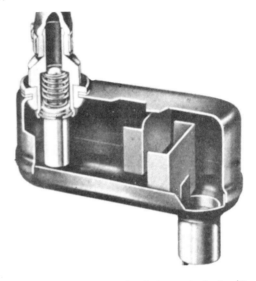

Fig. 1.17. The oil separator and emission control valve (Sec. 22)

Fig. 1.18. The crankcase emission control valve components (Sec. 22)

engine. Severe juddering when reversing or when moving off from rest is also likely and is a further sign of worn mounting rubbers.

2 The front mounting rubber insulators can be changed with the engine in the car.

3 Apply the handbrake firmly, jack up the front of the car, and place stands under the front of the car.

4 Lower the jack, take off the engine sump shield where fitted, and place the jack under the sump to take the weight of the engine using a square piece of wood to distribute the load.

5 Undo the large bolt which holds each of the engine mountings to the body crossmember. Then knock back the locking tabs and undo the four bolts holding each of the engine mountings in place.

6 Fit new mountings using new tab washers and tighten the four bolts down and bend up the locking tabs.

7 Screw in the bolts which connect the mountings to each side of the crossmember.

8 Remove the jack from under the engine and replace under the front crossmember, remove the stands and lower the car to the ground.

24 Examination and renovation - general

1 With the engine stripped down and all parts thoroughly cleaned, it is now time to examine everything for wear. The following items should be checked and where necessary renewed or renovated as described in the following Sections.

25 Crankshaft - examination and renovation

1 Examine the crankpin and main journal surfaces for signs of scoring or scratches. Check the ovality of the crankpins at different positions with a micrometer. If more than 0.001 inch (0.0254 mm) out of round, the crankpins will have to be reground. They will also have to be reground if there are any scores or scratches present. Also check the journals in the same fashion.

2 If it is necessary to regrind the crankshaft and fit new bearings, your local Ford garage or engineering works will be able to decide how much metal to grind off and the size of new bearing shells.

26 Big-end and main bearings - examination and renovation

1 Big-end bearing failure is accompanied by a knocking from the crankcase, and a slight drop in oil pressure. Main bearing failure is accompanied by vibration which can be quite severe as the engine speed rises. Inspect the big-end, main bearings, and thrust washers for signs of general wear, scoring, pitting and scratches. The bearings should be matt grey in colour. With lead indium bearings, should a trace of copper colour be noticed, the bearings are badly worn as the lead bearing material has worn away to expose the indium underlay. Renew the bearings if they are in this condition or if there is any sign of scoring or pitting.

2 The undersizes available are designed to correspond with the regrind sizes, ie. —0.010 inch (0.2540 mm) bearings are correct for a crankshaft reground —0.010 inch (0.2540 mm) undersize. The bearings are in fact slightly more than the stated undersize as running clearances have been allowed for during their manufacture.
3 Very long engine life can be achieved by changing big-end bearings at intervals of 30,000 miles (48,000 km) and main bearings at intervals of 50,000 miles (80,000 km), irrespective of bearing wear. Normally, crankshaft wear is infinitesimal and a change of bearings will ensure mileages of between 80,000 to 100,000 miles (128,000 to 161,000 km) before crankshaft regrinding becomes necessary. Crankshafts normally have to be reground because of scoring due to bearing failure.

27 Cylinder bores - examination and renovation

1 The cylinder bores must be examined for taper, ovality, scoring and scratches. Start by carefully examining the top of the cylinder bores. If they are at all worn, a very slight ridge will be found on the thrust side. This marks the top of the piston ring travel. The owner will have a good indication of the bore wear prior to dismantling the engine, or removing the cylinder head. Excessive oil consumption accompanied by blue smoke from the exhaust is a sure sign of worn cylinder bores and piston rings.
2 Measure the bore diameter just under the ridge with a micrometer and compare it with the diameter at the bottom of the bore which is not subject to wear. If the difference between the two measurements is more than 0.006 inch (0.1524 mm) it will be necessary to fit special pistons and rings or to have the cylinders rebored and fit oversize pistons. If a micrometer is not available, remove the rings from each piston in turn (do not mix the rings from piston to piston) and place each piston in its respective bore about ¾ inch (19.05 mm) below the top surface of the cylinder block. If a 0.010 inch (0.2540 mm) feeler gauge can be slid between the piston and the cylinder wall on the thrust side of the bore, then the following action must be taken. Oversize pistons are available (see Specifications Section at beginning of this Chapter).
3 These are accurately machined to just below these measurements so as to provide correct running clearances in bores of the exact oversize dimensions.
4 If the bores are slightly worn but not so badly as to justify reboring them, then special oil control rings and pistons can be fitted which will restore compression and stop the engine burning oil. Several different types are available and the manufacturers' instructions concerning their fitting must be followed closely.
5 If new pistons are being fitted and the bores have not been honed, it is essential to slightly roughen the hard glaze on the sides of the bores with fine glass paper so the new piston rings will have a chance to bed in properly.
6 Newly fitted pistons should be tested for clearance using a feeler gauge and spring balance. Place the feeler gauge between the piston and cylinder wall and having attached the spring balance to it, check the pull required to remove it is between 7 and 11 lb (3.2 and 4.5 kg) using a feeler blade ½ in wide (12.7 mm) and 0.0025 in (0.064 mm) thick.

28 Pistons and piston rings - examination and renovation

1 If the old pistons are to be refitted, carefully remove the piston rings and then thoroughly clean them. Take particular care to clean out the piston ring grooves. At the same time do not scratch the aluminium in any way. If new rings are to be fitted to the old pistons then the top ring should be stepped so as to clear the ridge left above the previous top ring. If a normal but oversize new ring is fitted, it will hit the ridge and break because the new ring will not have worn in the same way as the old. This will have worn in unison with the ridge.
2 Before fitting the rings on the pistons, each should be inserted approximately 2 inches (50 mm) down the cylinder bore and the end gap measured with a feeler gauge. This should be between 0.009 inch (0.2286 mm) and 0.014 inch (90.3556 mm). It is essential that the gap should not be measured at the top of a worn bore, and although giving a perfect fit, it could easily seize at the bottom. If the ring gap is too small rub down the ends of the ring with a very fine file until the gap, when fitted, is correct. To keep the rings square in the bore for measurement, line each up in turn by inserting an old piston in the bore upside down, and use the piston to push the ring down about 2 inches (50 mm). Remove the piston and measure the piston ring end gap.
3 When fitting new pistons and rings to a rebored engine, the piston ring end gap can be measured at the top of the bore as the bore will not now taper. It is not necessary to measure the side clearance in the piston ring grooves with the rings fitted as the groove dimensions are accurately machined during manufacture. When fitting new oil control rings to old pistons, it may be necessary to have the grooves widened by machining to accept the new wider rings. In this instance the manufacturer's fitting instructions will indicate the procedure.

29 Camshaft and camshaft bearings - examination and renovation

1 Carefully examine the camshaft bearings for wear. If the bearings are obviously worn or pitted, then they must be renewed. This is an operation for your local Ford dealer or engine reconditioning works as it demands the use of specialised equipment. The bearings are removed with a special drift after which new bearings are pressed in, care being taken to ensure the oil holes in the bearing line up with those in the block.
2 The camshaft itself should show no signs of wear. If scoring on the cams is noticed, the only permanently satisfactory cure is to fit a new camshaft.
3 Examine the skew gear for wear, chipped teeth or other damage.
4 Carefully examine the camshaft thrust plate. Excessive wear will be visually self-evident and will require the fitting of a new plate.

30 Valves and valve seats - examination and renovation

1 Examine the heads of the valves for pitting and burning, especially the heads of the exhaust valves. The valve seatings should be examined at the same time. If the pitting on valve and seat is very slight, the marks can be removed by grinding the exhaust seats and valves together with coarse, and then fine, valve grinding paste.
2 Where bad pitting has occurred to the valve seats, it will be necessary to recut them and fit new valves. If the valve seats are so worn that they cannot be recut, then it will be necessary to fit new valve seat inserts. These latter two jobs should be entrusted to the local Ford agent or engineering works. In practice it is very seldom that the seats are so badly worn that they require renewal. Normally, it is the valve that is too badly worn for replacement, and the owner can easily purchase a new set of valves and match them to the seats by valve grinding.
3 Valve grinding is carried out as follows: Smear a trace of coarse carborundum paste on the seat face and apply a suction grinder tool to the valve head. With a semi-rotary motion, grind the valve head to its seat, lifting the valve occasionally to redistribute the grinding paste. When a dull matt even surface finish is produced on both the valve seat and the valve, then wipe off the paste and repeat the process with fine carborundum paste, lifting and turning the valve to redistribute the paste as before. A light spring placed under the valve head will greatly ease this operation. When a smooth, unbroken ring of light

Fig. 1.19. Use a suction tool to hold the valve head while grinding-in (Sec. 30)

grey matt finish is produced, on both valve and valve seat faces, the grinding operation is complete.
4   Scrape away all carbon from the valve head and the valve stem. Carefully clean away every trace of grinding compound, taking great care to leave none in the ports or in the valve guides. Clean the valves and valve seats with a paraffin soaked rag then with a clean rag, and finally, if an air line is available, blow the valves, valve guides and valve ports clean.

31 Timing sprockets and chain - examination and renovation

1   Examine the teeth on both the crankshaft sprocket and the camshaft sprocket for wear. Each tooth forms an inverted V with the sprocket periphery, and if worn, the side of each tooth under tension will be slightly concave in shape when compared with the other side of the tooth, ie. one side of the inverted V will be concave when compared with the other. If any sign of wear is present the sprockets must be renewed.
2   Examine the links of the chain for side slackness and renew the chain if any slackness is noticeable when compared with a new chain. It is a sensible precaution to renew the chain at about 30,000 miles (48,000 km) and at a lesser mileage if the engine is stripped down for a major overhaul. The rollers on a very badly worn chain may be slightly grooved.

32 Rockers and rocker shaft - examination and renovation

1   Thoroughly clean the rocker shaft and then check it for distortion by rolling it on a piece of plate glass. If it is out of true, renew it. The surface of the shaft should be free from wear ridges and score marks.
2   Check the rocker arms for wear of the rocker bushes, for wear at the rocker arm face which bears on the valve stem, and for wear of the adjusting ball ended screws. Wear in the rocker arm bush can be checked by gripping the rocker arm tip and holding the rocker arm in place on the shaft, noting if there is any lateral rocker arm shake. If shake is present, and the arm is very loose on the shaft, a new bush or rocker arm must be fitted.
3   Check the top of the rocker arm where it bears on the valve head for cracking or serious wear on the case hardening. If none is present re-use the rocker arm. Check the lower half of the ball on the end of the rocker arm adjusting screw. Check the pushrods for straightness by rolling them on a piece of plate glass. Renew any that are bent.

33 Tappets (cam followers) - examination and renovation

Examine the bearing surface of the mushroom tappets which lie on the camshaft. Any indentation in this surface or any cracks indicate serious wear and the tappets should be renewed. Thoroughly clean them out, removing all traces of sludge. It is most unlikely that the sides of the tappets will prove worn, but if they are a very loose fit in their bores and can readily be rocked, they should be exchanged for new units. It is very unusual to find any wear in the tappets, and any wear is likely to occur only at very high mileages.

34 Connecting rods - examination and renovation

1   Examine the mating surfaces of the big-end caps to see if they have ever been filed in a mistaken attempt to take up wear. If so the offending rods must be renewed.
2   Insert the gudgeon pin into the little end of the connecting rod. It should go in fairly easily, but if any slackness is present, then take the rod to your local Ford dealer and exchange it for a rod of identical weight.
3   When the pistons have been reassembled to the rods, it is a good system to have the rod alignment checked by your local Ford dealer or engine reconditioners prior to refitting to the cylinder block.

35 Starter ring gear - examination and renovation

1   If the flywheel ring gear teeth are worn then the ring gear can be renewed without the need to replace the flywheel.
2   To remove a starter ring either split it with a cold chisel after making a cut with a hacksaw blade between two teeth, or heat the ring, and use a soft-headed hammer (not steel) to knock the ring off, striking it evenly and alternately, at equally spaced points. Take great care not to damage the flywheel during this process.
3   Clean and polish with emery cloth four evenly spaced areas on the outside face of the new starter ring.
4   Heat the ring evenly with an oxyacetylene flame until the polished portions turn dark blue (600°F/316°C). Hold the ring at this temperature for five minutes and then quickly fit it to the flywheel so the chamfered portion of the teeth faces the gearbox side of the flywheel.
5   The ring should be tapped gently down onto its register and left to cool naturally when the contraction of the metal on cooling will ensure that it is a secure and permanent fit. Great care must be taken not to overheat the ring, indicated by it turning light metallic blue, as if this happens the temper of the ring will be lost.
6   It does not matter which way round the ring for pre-engaged starters is fitted as it has no chamfers on its teeth. This also makes for quick identification between the two rings.

36 Cylinder head - decarbonising

1   This can be carried out with the engine either in or out of the car. With the cylinder head off, carefully remove, with a wire brush mounted in an electric drill and a blunt scraper, all traces of carbon deposits from the combustion spaces and the ports. The valve head, stems and valve guides should also be freed from any carbon deposits. Wash the combustion spaces and ports down with petrol and scrape the cylinder head surface free of any foreign matter with the side of a steel rule, or a similar article.
2   Clean the pistons and top of the cylinder bores. If the pistons are still in the block then it is essential that great care is taken to ensure that no carbon gets into the cylinder bores as this could scratch the cylinder walls or cause damage to the piston and rings. To ensure that this does not happen, first turn the crankshaft so that two of the pistons are at the top of their bores. Stuff rag into the other two bores or seal them off with paper and masking tape. The waterways should also be covered with small pieces of masking tape to prevent particles of carbon entering the cooling system and damaging the water pump.
3   There are two schools of thought as to how much carbon should be removed from the piston crown. One school recommends that a ring of carbon should be left round the edge of the piston and on the cylinder bore wall as an aid to low oil consumption. Although this is probably true for early engines with worn bores, on modern engines it is preferable to remove all traces of carbon deposits.
4   If all traces of carbon are to be removed, press a little grease into the gap between the cylinder walls and the two pistons which are to be worked on. With a blunt scraper carefully scrape all the carbon from the piston crown, taking great care not to scratch the aluminium. Also scrape away the carbon from the surrounding lip of the cylinder wall. When all carbon has been removed, scrape away the grease which will now be contaminated with carbon particles, taking care not to press any into the bores. To assist prevention of carbon build-up the piston crown can be polished with a metal polish. Remove the rags or masking tape from the other two cylinders and turn the crankshaft so that the two pistons which were at the bottom are now at the top. Place rag or masking tape in the cylinders which have been decarbonised and proceed as already described.
5   Thoroughly clean out the cylinder head bolt holes in the top face of the block. If these are filled with carbon, oil or water it is possible for the block to crack when the bolts are screwed in due to the hydraulic pressure created by the trapped fluid.

37 Valve guides - examination and renovation

1   Examine the valve guides internally for scoring and other signs of wear. If a new valve is a very loose fit in a guide and there is a trace of lateral rocking then new guides will have to be fitted.
2   The fitting of new guides is a job which should be done by your local Ford dealer.

38 Engine reassembly - general

1 To ensure maximum life with minimum trouble from a rebuilt engine, not only must everything be correctly assembled, but everything must be spotlessly clean, all the oilways must be clear, locking washers and spring washers must always be fitted where indicated and all bearing and other working surfaces must be thoroughly lubricated during assembly.
2 Before assembly begins renew any bolts or studs, the threads of which are in any way damaged, and whenever possible use new spring washers.
3 Apart from your normal tools, a supply of clean rag, an oil can filled with engine oil, a new supply of assorted spring washers, a set of new gaskets, and a torque wrench, should be collected together.

39 Crankshaft and rear oilseal - refitting

1 Thoroughly clean the block and ensure that all traces of old gaskets etc are removed.
2 Position the upper halves of the shell bearings in their correct positions so that the tabs of the shells engage in the machined keyways in the sides of the bearing locations (photo).
3 Oil the main bearing shells after they have been fitted in position (photo).
4 Thoroughly clean out the oilways in the crankshaft with the aid of a thin wire and blow clean with an air line.
5 To check for the possibility of an error in the grinding of the crankshaft journal (presuming the crankshaft has been reground) smear engineers blue evenly over each big-end journal in turn with the crankshaft end flange held firmly in position in a vice.
6 With new shell bearings fitted to the connecting rods fit the correct rod to each journal in turn, fully tightening down the securing bolts.
7 Spin the rod on the crankshaft a few times and then remove the big-end cap. A fine unbroken layer of engineers blue should cover the whole of the journal. If the blue is much darker on one side than the other or if the blue has disappeared from a certain area (ignore the very edges of the journal) then something is wrong and the journal will have to be checked with a micrometer.
8 The main journals should also be checked in a similar fashion with the crankshaft in the crankcase. On completion of these tests remove all traces of the engineers blue.
9 The crankshaft can now be lowered carefully into place (photo).
10 Fit new endfloat thrust washers. These locate in recesses on each side of the centre main bearing in the cylinder block and must be fitted with oil grooves facing the crankshaft flange. With the crankshaft in position check for endfloat which should be between 0.003 inch (0.075 mm) and 0.011 inch (0.279 mm). If the endfloat is correct, remove the thrust washer and select suitable washers to give the correct endfloat.
11 Place the lower halves of the main bearing shells in their caps, making sure that the locking tabs fit into the machined grooves. Refit the main bearing caps ensuring that they are the correct way round and that the correct cap is on the correct journal. The front cap is marked 'F', the second 'R2', the centre cap 'C', the fifth cap 'RA' and the rear cap 'R'. The indicating arrow must point to the front on each cap. Tighten the cap bolts to a torque of 55-60 lb f ft (7.5-8.2 kg f m) (photo). Spin the crankshaft to make certain it is turning freely.
12 Fit a new rear main oilseal bearing retainer gasket to the rear of the cylinder block (photo).
13 Then fit the rear main oilseal bearing retainer housing (photo). Note that the oilseal is also circular and is simply prised out when removed, a new one being pressed in (Fig. 1.20).
14 Lightly tighten the four retaining bolts with spring washers under their heads noting that two bolts are dowelled to ensure correct alignment and should be tightened first.
15 Torque the bolts to 12 to 15 lb f ft (1.66 to 2.07 kg f m) and check that the housing is centralised.

40 Pistons and connecting rods - refitting

1 Check that the piston ring grooves and oilways are thoroughly clean

39.2 Upper half of main bearing shell in place on cylinder block. Note shell tab engaged in block keyway (arrowed)

39.3 Oiling the main bearing shells

39.9 Refitting the crankshaft

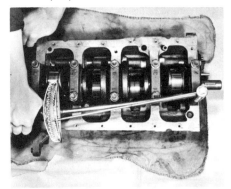

39.11 Tighten the main bearing cap bolts to correct torque

39.12 Fit the oil seal housing gasket to the rear of the cylinder block ...

39.13 ... then fit the housing

Chapter 1/Engine

Fig. 1.20. Installing an oil seal to a circular crankshaft rear retainer (Sec. 39)

hole in the piston and fit the piston to the connecting rod by sliding in the gudgeon pin. The arrow on the crown of each piston must be on the same side as the word 'FRONT' on the connecting rod (photos).

5 Fit the second circlip in position. Repeat this procedure for the remaining three pistons and connecting rods.

6 Fit the connecting rod in position and check that the oil hole in the upper half of each bearing aligns with the oil squirt hole in the connecting rod (photo).

7 With a wad of clean rag wipe the cylinder bores clean, and then oil them generously. The pistons complete with connecting rods, are fitted to their bores from above. As each piston is inserted into its bore, ensure that it is the correct piston/connecting rod assembly for that particular bore and that the connecting rod is the right way round, and that the front of the piston is towards the front of the bore, ie. towards the front of the engine.

8 The piston will only slide into the bore as far as the oil control ring. It is then necessary to compress the piston rings in a clamp (photos).

9 Gently tap the piston into the cylinder bore with a wooden or plastic hammer (photo). If a proper piston ring clamp is not available

and unblocked. Piston rings must always be fitted over the head of the piston and never from the bottom. Fit the rings by the same method used for removing them (Section 15).

2 When assembling the rings note that the compression rings are marked 'top', and that the upper ring is chromium plated. The ring gaps should be spaced at 120° angles round the piston (photos).

3 If the same pistons are being re-used, then they must be mated to the same connecting rod with the same gudgeon pin. If new pistons are being fitted it does not matter which connecting rod they are used with. Note that the word 'FRONT' is stamped on one side of each of the rods. On reassembly the side marked 'FRONT' must be towards the front of the engine (photo).

4 Fit a gudgeon pin circlip in position at one end of the gudgeon pin

Fig. 1.21. Relationship between piston (arrow) and conrod (front) (Sec. 40)

40.2a 'Top' markings on piston compression rings

40.2b Piston rings in place on piston with gaps set at 120° angles around the piston

40.3 The word 'front' on the connecting rod must be towards the front of the engine on reassembly

40.4a Replacing the gudgeon pin circlip prior to inserting gudgeon pin and conrod

40.4b Inserting gudgeon pin through piston and conrod

40.6 Fitting connecting rod upper shell in conrod. Note alignment of shell and conrod oil holes

40.8a Using a piston ring clamp to compress rings ...

40.8b ... prior to inserting piston in bore

40.9 Using the shaft of a hammer to tap piston down the bore

40.12 Torqueing connecting rod bearing caps

40.13 Semi rebuilt engine ready to receive camshaft and tappets

41.1a Inserting the cam followers into their bores

41.1b The cam followers, inserted into their bores

41.2 Carefully insert the camshaft

41.3a Inserting the thrust plate behind the camshaft

41.3b Plate bolted in position with tab washers turned up to lock bolt. Note timing gear locating peg (arrowed)

41.6 Crankshaft and camshaft timing sprocket marks aligned with peg fully inserted in camshaft sprocket (arrows)

41.8 Bending up tabs of lockwasher to retain camshaft sprocket bolts

Chapter 1/Engine

then a suitable jubilee clip does the job very well.
10 Note the directional arrow on the piston crown (Fig. 1.21).
11 Fit the shell bearings to the big-end caps so the tongue on the back of each bearing lies in the machined recess.
12 Generously oil the crankshaft connecting rod journals and then replace each big-end cap on the same connecting rod from which it was removed. Fit the locking plates under the heads of the big-end bolts, tap the caps right home on the dowels and then tighten the bolts to a torque of 31-35 lb f ft (4.2-4.8 kg f m). To facilitate reassembly the rod and cap are marked (ie. 1 - 2 - 3 - 4) these numbers should be together and on the camshaft side of the engine (photo).
13 The semi-rebuilt engine is now ready for the cam followers and cam to be fitted (photo).

41 Camshaft and cam followers - refitting

1 Lubricate and fit the eight cam followers into the same holes in the block from which each was removed. The cam followers can only be fitted with the block upside down (photos).
2 Fit the Woodruff key in its slot on the front of the crankshaft and then press the timing sprocket into place so the timing mark faces forward. Oil the camshaft shell bearings and insert the camshaft into the block (which should still be upside down) (photo).
3 Make sure the camshaft turns freely and then fit the thrust plate behind the camshaft flange as shown in the photograph. Torque the thrust plate bolts to 2.5 to 3.5 lb f ft (0.35 to 0.48 kg f m). Measure the endfloat with a feeler gauge - it should be between 0.0025 and 0.0075 inch (0.0635 and 0.1905 mm). If this is not so, then renew the plate (photo).
4 Turn up the tab under the head of each bolt to lock it in place.
5 Refit the camshaft timing gear and loosely retain with its two retaining bolts. Use a **new** tab washer.

6 When refitting the timing chain round the gearwheels and to the engine, the two timing lines (arrowed) must be adjacent to each other on an imaginary line passing through each gearwheel centre (photo).
7 With the timing marks correctly aligned turn the camshaft until the protruding dowel locates in the hole (arrowed) in the camshaft sprocket wheel.
8 Tighten the two retaining bolts and bend up the tabs on the lockwasher (photo).

42 Timing chain tensioner and cover - refitting

1 Fit the oil slinger to the nose of the crankshaft, concave side facing outwards. The cut-out locates over the Woodruff key.
2 Then slide the timing chain tensioner arm over its hinge pin on the front of the block.
3 Turn the tensioner back from its free position so that it will apply pressure to the tensioner arm and replace the tensioner on the block sump flange.
4 Bolt the tensioner to the block using spring washers under the heads of the two bolts (photo).
5 Remove the front oilseal from the timing chain cover and carefully press a new seal into position (photo). Lightly lubricate the face of the seal which will bear against the crankshaft. If available, borrow the special tool from your Ford dealer to assist in accurate alignment when refitting the new oilseal, see Fig. 1.22.
6 Using jointing compound, fit a new timing cover gasket in place (photo).
7 Fit the timing chain cover, replacing and tightening the two dowel bolts first. These fit in the holes nearest the sump flange and serve to align the timing cover correctly. Ensure spring washers are used and then tighten the bolts evenly.

42.4 Timing chain tensioner in place on underside of cylinder block

42.5 Replacing timing cover oil seal

42.6 Timing cover gasket in position on block ready to receive timing cover

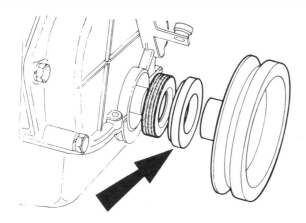
Fig. 1.22. Special tool (arrowed) available from Ford dealers to assist in accurate alignment of front cover oil seal during replacement (Sec. 42)

Fig. 1.23. Replacing the oil slinger and timing cover complete with oil seal (Sec. 42)

43 Flywheel, sump, oil pump and crankshaft pulley - refitting

1 Refit the tube or crankcase emission device to its recess adjacent to the top of the petrol pump, tapping it gently into place (photo). Replace the oil pump suction pipe using a new tab washer and position the gauze head so that it clears the crankshaft throw and the oil return pipe (where fitted). Tighten the nut and bend back the tab of the lockwasher.

2 Clean the flanges of the sump and fit new gaskets in place. Fit a new oil seal to the flange at the rear of the crankcase and at the front (photos). Ensure that the sump gasket end tabs locate under the ends of the rear seal.

3 With the rear cover in position, locate the flywheel onto the crankshaft flange and tighten the securing bolts to a torque of between 50 to 56 lb f ft (6.8 to 7.6 kg fm) (photo).

4 Locate the sump in position on the crankcase and tighten the securing bolts evenly in diagonal sequence.

5 The engine can now be turned over so that it is the right way up. Coat the oil pump flanges with jointing compound.

6 Fit a new gasket in place on the oil pump.

7 Position the oil pump against the block ensuring that the skew gear teeth on the driveshaft mate with those on the camshaft (photo).

8 Replace the three securing bolts and spring washers and tighten them down evenly.

9 Moving to the front of the engine align the slot in the crankshaft pulley wheel with the key on the crankshaft and gently tap the pulley wheel home.

10 Secure the pulley wheel by fitting the large flat washer, the spring washer and then the bolt which should be tightened securely (photo).

44 Cylinder head and valves - reassembly and refitting

1 Thoroughly clean the faces of the block and cylinder head. Then fit a new cylinder head gasket. In order to correctly position the gasket it

43.1 Oil pick-up pipe bracket location on underside of cylinder block

43.2a Sump gasket in position on cylinder block sump flange

43.2b Oil seals being fitted in timing cover ...

43.2c ... and engine rear oil seal carrier

43.3 Torqueing flywheel retaining bolts

43.7 Oil pump in place on cylinder block prior to receiving retaining bolts

43.10 Tightening the crankshaft pulley bolt

44.3 Valve stem oil seal correctly fitted

44.7 Compress the spring and fit the collets

is a good idea to temporarily screw in two lengths of studding (one in each extreme diagonal hole) to act as locating dowels. These should be removed once two of the cylinder head bolts have been screwed into position.

2 With the cylinder head on its side lubricate the valve stems and refit the valves to their correct guides. The valves should previously have been ground in (see Section 30).

3 Then fit the valve stem umbrella oil seals open ends down (photo).

4 Next slide the valve spring into place. Use new ones if the old set has covered 20,000 miles (32,000 km).

5 Slide the valve spring retainer over the valve stem.

6 Compress the valve spring with a compressor.

7 Then refit the split collets (photo). A trace of grease will help to hold them to the valve stem recess until the spring compressor is slackened off and the collets are wedged in place by the spring cap.

8 Carefully lower the cylinder head onto the block (Fig. 1.24).

9 Replace the cylinder head bolts, cleaned and oiled, and screw them down finger tight. Note that two of the bolts are of a different length.

10 With a torque wrench tighten the bolts to 66 to 71 lb f ft (9.0 to 9.7 kg fm), in the order shown in Fig. 1.25.

45 Rocker shaft and push rods - refitting

1 Lubricate and reassemble the various rocker shaft components in the reverse order to dismantling.

2 Refit the push rods into their respective holes in the block, and ensure that each rod is seated correctly in the cam follower.

3 Now refit the rocker shaft to the cylinder head. Ensure that the oil holes are clear and that the cut outs for the securing bolts lie facing the holes in the brackets.

4 Check that each rocker arm and push rod are correctly located (photo), and with the four rocker pillar bolts and washers located, tighten to a torque of 18 to 22 lb f ft (2.4 to 3 kg fm) (photo).

46 Rocker arm/valve - adjustment

1 The valve adjustments should be made with the engine cold. The importance of correct rocker arm/valve stem clearances cannot be overstressed as they vitally affect the performance of the engine. If the clearances are set too open, the efficiency of the engine is reduced as the valves open late and close earlier than was intended. If, on the other hand, the clearances are set too close there is a danger that the stems will expand upon heating and not allow the valves to close properly which will cause burning of the valve head and seat and possible warping. If the engine is in the car access to the rockers is by removing the four holding down screws from the rocker cover, and then lifting the rocker cover and gasket away.

2 It is important that the clearance is set when the tappet of the valve being adjusted is on the heel of the cam (ie. opposite the peak). This can be ensured by carrying out the adjustments in the following order (which also avoids turning the crankshaft more than necessary):

Valves open	Valves to adjust
1 ex 6 in	3 in 8 ex
2 in 4 ex	5 ex 7 in
3 in 8 ex	1 ex 6 in
5 ex 7 in	2 in 4 ex

Fig. 1.24. Replacement of the cylinder head is made easier if two old bolts are ground down and placed as at 'A'

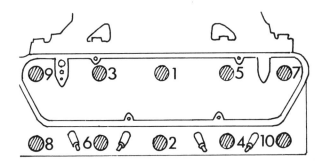

Fig. 1.25. Cylinder head bolt tightening sequence: Torque bolts in order shown (Sec. 8 and 44)

45.4a Ensure that the rocker shaft adjusting nuts engage their respective pushrods before tightening the rocker shaft bolts

45.4b Torqueing the rocker shaft bolts

The valve positions are numbered from the front of the engine, and the clearances are given in the Specifications.

3 Working from the front of the engine (No.1 valve), the correct clearance is obtained by rotating the hexagon adjuster with a spanner whilst the appropriate size feeler gauge is installed between the valve stem and the rocker arm. The feeler gauge should be a firm sliding fit; take care that it is not pinched by over-tightening the adjuster (photo).

4 Do not refit the rocker cover before replacing the distributor and setting the ignition timing.

47 Distributor - refitting

1 It is important to set the distributor drive correctly as otherwise the ignition timing will be totally incorrect.

2 It is possible to set the distributor drive in apparently the right position, but, in fact, 180° out by omitting to select the correct cylinder which must not only be at TDC but must also be on its firing stroke with both valves closed. The distributor drive should therefore not be fitted until the cylinder head is in position and the valves can be observed. Alternatively, if the timing cover has not been replaced, the distributor drive can be replaced when the lines on the timing wheels are adjacent to each other.

3 Rotate the crankshaft so that No 1 piston is at TDC and on its firing stroke (the lines in the timing gears will be adjacent to each other). When No 1 piston is at TDC both valves will be closed and both rocker arms will 'rock' slightly because of the stem to arm pad clearance.

4 Note the timing marks on the timing case and the notch on the crankshaft wheel periphery (photo). Set the crankshaft so the cut-out is in the right position of initial advance which is 6° BTDC.

5 Hold the distributor in place so that the vacuum unit is towards the rear of the engine and at an angle of about 30° to the block. Do not yet engage the distributor drive gear with the skew gear on the camshaft.

6 Turn the rotor arm so that it points toward No 2 inlet port (photo).

7 Push the distributor shaft into its bore and note, as the distributor drive gear and skew gear on the camshaft mate, that the rotor arm turns so that it assumes a position of approximately 90° to the engine (photo). Fit the bolt and washer which holds the distributor clamp plate to the block.

8 Loosen the clamp on the base of the distributor and slightly turn the distributor body until the points just start to open while holding the rotor arm against the direction of rotation so no lost motion is present. Tighten the clamp. For a full description of how to do this accurately see Chapter 4.

48 Engine - final assembly

1 Reconnect the ancillary components to the engine in the reverse order to which they were removed.

2 Fit a new gasket to the water pump and attach the pump to the front of the cylinder block (photo). Note that the generator adjustment strap fits under the head of the lower bolt on the water pump.

3 Replace the fuel pump using a new gasket and tighten up the two securing bolts.

4 Fit the thermostat and thermostat gasket to the cylinder head and then replace the thermostat outlet pipe. Replace the spark plugs and refit the rocker cover using a new gasket.

5 Refit the generator and adjust it so there is ½ inch (12.7 mm) play in the fan belt between the water pump and generator pulley. Refit the vacuum advance pipe to the distributor and refit the sender units.

6 Refit the inlet and exhaust manifolds together with the carburettor using new gaskets.

7 The generator, thermostat, oil filter and engine mounting brackets should also be reassembled prior to refitting the engine to the car.

46.3 Using feeler gauge and ring spanner to adjust the valve clearances

47.4 Timing marks cast into timing cover and cut into crankshaft pulley flange (arrowed)

47.6 Position of rotor arm before installing distributor

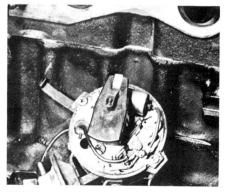

47.7 Position of rotor arm after installing distributor

48.2 Refitting the water pump

49 Engine - installation (general)

Although the engine can be installed by one man with a suitable winch, it is easier if two are present: one to lower the engine into the engine compartment and the other to guide the engine into position and to ensure that it does not foul anything.

At this stage one or two tips may come in useful. Ensure all the loose leads, cables, etc are tucked out of the way. If not, it is easy to trap one and so cause much additional work after the engine is replaced. Smear grease on the top of the gearbox input shaft before fitting the gearbox.

Always fit a new fan belt and new cooling hoses and jubilee clips as this will help eliminate the possibility of failure while on the road.

50 Engine - installation (without gearbox)

1 Position a sling round the engine and secure it to the hoist.
2 Lower the engine into the engine compartment, ensuring that nothing is fouling. Align the height of the engine with the gearbox which will, it is assumed, still be supported on the jack which was located prior to removal of the engine.
3 Move the engine rearward until the splines of the gearbox first motion shaft enter the splined hub of the clutch driven plate (friction disc). The clutch driven plate will have already been aligned as described in Chapter 5. The engine may need turning fractionally to obtain engagement. If so, turn the crankshaft pulley using a spanner applied to its centre bolt.
4 Move the engine fully to the rear to mate the faces of the clutch bellhousing and the engine crankcase. Insert and tighten the securing bolts, via the engine mountings and clutch housing.
5 Reconnect the engine front mountings and bolt them to the body sideframe members (it will be easier if the weight of the engine is still taken by the hoist so that it can be moved slightly to align the mounting bolt holes).
6 Connect the starter motor cable and the engine earth strap.
7 Reconnect the servo vacuum pipe to the inlet manifold.
8 Unplug the fuel line and connect it to the fuel pump.
9 Reconnect coil, distributor and spark plug leads.
10 Connect the exhaust downpipe and connect the accelerator cable bracket and choke controls.
11 Connect the oil pressure switch and temperature gauge transmitter unit leads.
12 Fit the radiator together with radiator and heater hoses, followed by the engine shield and the air cleaner (do not overtighten the air cleaner bolts or the carburettor may fracture).
13 Connect the leads to the rear of the alternator and then connect the battery negative terminal.
14 Refill the cooling system (Chapter 2).
15 Refill the engine with the correct grade and quantity of oil (an extra pint will be required for absorption by the new filter element).
16 Refit the bonnet and check its alignment before tightening the hinge bolts fully. This operation will be easier to perform if the help of an assistant is obtained.

51 Engine - installation (with gearbox)

1 Position a sling round the engine/gearbox unit and support its weight on suitable lifting tackle. If using a fixed hoist raise the power unit and roll the car under it so the power unit will easily drop into the engine compartment.
2 Lower the power unit into position moving the car forward at the same time. When the engine is three-quarters in it will be found helpful to place a trolley jack under the gearbox.
3 Connect the engine front mountings and the rear one with its supporting crossmember.
4 Fit the propeller shaft, aligning the drive flange mating marks made before removal.
5 Connect the clutch operating cable and adjust the clutch, as described in Chapter 5.
6 Refit the gearlever and gaiter.
7 Carry out operations 6 to 16 as described in the preceding Section.

52 Engine - initial start up after major overhaul

1 There is no reason why the reassembled engine should not fire at the first operation of the starter switch.
2 If it fails to do so, make two or three more attempts as it may be that the carburettor bowl is empty and requires filling by a few revolutions of the camshaft operating fuel pump.
3 If the engine still does not fire, check the following points:
   a) *There is fuel in the tank.*
   b) *Ignition and battery leads are correctly and securely connected. (Check particularly the spark plug HT lead sequence - Chapter 4).*
   c) *The choke is correctly connected.*
   d) *The distributor has been correctly installed and not fitted 180° out (Section 47).*
   e) *Work systematically through the fault diagnosis chart at the end of this Chapter.*
4 Run the engine until normal operating temperature is reached and check the torque setting of all nuts and bolts, particularly the cylinder head bolts. This is done by slackening the bolts slightly and retightening to the correct torque.
5 Adjust the slow-running and carburettor mixture control screws (Chapter 3).
6 Check for any oil or water leaks and when the engine has cooled, check the levels of the radiator and sump and top-up as necessary.

53 Fault diagnosis - engine

Symptom	Reason/s	Remedy
Engine fails to turn over when starter operated		
No current at starter motor	Flat or defective battery	Charge or replace battery. Push-start car.
	Loose battery leads	Tighten both terminals and earth ends of earth lead.
	Defective starter solenoid or switch or broken wiring	Run a wire direct from the battery to the starter motor or by-pass the solenoid.
	Engine earth strap disconnected	Check and retighten strap.
Current at starter motor	Jammed starter motor drive pinion	Place car in gear and rock from side to side. Alternatively, free exposed square end of shaft with spanner.
	Defective starter motor	Remove and recondition.

Chapter 1/Engine

Symptom	Reason/s	Remedy
Engine turns over but will not start No spark at spark plug	Ignition damp or wet	Wipe dry the distributor cap and ignition leads.
	Ignition leads to spark plugs loose	Check and tighten at both spark plug and distributor cap ends.
	Shorted or disconnected low tension leads	Check the wiring on the CB and SW terminals of the coil and to the distributor.
	Dirty, incorrectly set, or pitted contact breaker points	Clean, file smooth, and adjust.
	Faulty condenser	Check contact breaker points for arcing, remove and fit new.
	Defective ignition switch	By-pass switch with wire.
	Ignition leads connected wrong way round	Remove and replace leads to spark plug in correct order.
	Faulty coil	Remove and fit new coil.
	Contact breaker point spring earthed or broken	Check spring is not touching metal part of distributor. Check insulator washers are correctly placed. Renew points if the spring is broken.
Excess of petrol in cylinder or carburettor flooding	Too much choke allowing too rich a mixture to wet plugs	Remove and dry spark plugs or with wide open throttle, push-start the car.
	Float damaged or leaking or needle not seating	Remove, examine, clean and replace float and needle valve as necessary.
	Float lever incorrectly adjusted	Remove and adjust correctly.
Engine stalls and will not start No spark at spark plug	Ignition failure - sudden	Check over low and high tension circuits for breaks in wiring.
	Ignition failure - misfiring precludes total stoppage	Check contact breaker points, clean and adjust. Renew condenser if faulty.
	Ignition failure - in severe rain or after traversing water splash	Dry out ignition leads and distributor cap.
No fuel at jets	No petrol in petrol tank	Refill tank.
	Petrol tank breather choked	Remove petrol cap and clean out breather hole or pipe.
	Sudden obstruction in carburettor(s)	Check jets, filter, and needle valve in float chamber for blockage.
	Water in fuel system	Drain tank and blow out fuel lines.
Engine misfires or idles unevenly Intermittent spark at spark plug	Ignition leads loose	Check and tighten as necessary at spark plug and distributor cap ends.
	Battery leads loose on terminals	Check and tighten terminal leads.
	Battery earth strap loose on body attachment point	Check and tighten earth lead to body attachment point.
	Engine earth lead loose	Tighten lead.
	Low tension leads to SW and CB terminals on coil loose	Check and tighten leads if found loose.
	Low tension lead from CB terminal side to distributor loose	Check and tighten if found loose.
	Dirty, or incorrectly gapped plugs	Remove, clean, and regap.
	Dirty, incorrectly set, or pitted contact breaker points	Clean, file smooth, and adjust.
	Tracking across inside of distributor cover	Remove and fit new cover.
	Ignition too retarded	Check and adjust ignition timing.
	Faulty coil	Remove and fit new coil.
No fuel at carburettor float chamber or at jets	No petrol in petrol tank	Refill tank.
	Vapour lock in fuel line (in hot conditions or at high altitude)	Blow into petrol tank, allow engine to cool or apply a cold wet rag to the fuel line.
	Blocked float chamber needle valve	Remove, clean and replace.
	Fuel pump filter blocked	Remove, clean, and replace.
	Choked or blocked carburettor jets	Dismantle and clean.
	Faulty fuel pump	Remove, overhaul and replace.
Fuel shortage at engine	Mixture too weak	Check jets, float chamber needle valve, and filters for obstruction. Clean as necessary. Carburettor incorrectly adjusted.
	Air leak in carburettor	Remove and overhaul carburettor.
	Air leak at inlet manifold to cylinder head, or inlet manifold to carburettor	Test by pouring oil along joints. Bubbles indicate leak. Renew manifold gasket as appropriate.

Chapter 1/Engine

Symptom	Reason/s	Remedy
Mechanical wear	Incorrect valve clearances	Adjust rocker arms to take up wear.
	Burnt out exhaust valves	Remove cylinder head and renew defective valves.
	Sticking or leaking valves	Remove cylinder head, clean, check and renew valves as necessary.
	Weak or broken valve springs	Check and renew as necessary.
	Worn valve guides or stems	Renew valve guides and valves.
	Worn pistons and piston rings	Dismantle engine, renew pistons and rings.
Lack of power and poor compression		
Fuel/air mixture leaking from cylinder	Burnt out exhaust valves	Remove cylinder head, renew defective valves.
	Sticking or leaking valves	Remove cylinder head, clean, check, and renew valves as necessary.
	Worn valve guides and stems	Remove cylinder head and renew valves and valve guides.
	Weak or broken valve springs	Remove cylinder head, renew defective springs.
	Blown cylinder head gasket (accompanied by increase in noise)	Remove cylinder head and fit new gasket.
	Worn pistons and piston rings	Dismantle engine, renew pistons and rings.
	Worn or scored cylinder bores	Dismantle engine, rebore, renew pistons and rings.
Incorrect adjustments	Ignition timing wrongly set. Too advanced or retarded	Check and reset ignition timing.
	Contact breaker points incorrectly gapped	Check and reset contact breaker points.
	Incorrect valve clearances	Check and reset rocker arm to valve stem gap.
	Incorrectly set spark plugs	Remove, clean and regap.
	Carburation too rich or too weak	Tune carburettor for optimum performance.
Carburation and ignition faults	Dirty contact breaker points	Remove, clean and replace.
	Distributor automatic balance weights or vacuum advance and retard mechanism not functioning correctly	Overhaul distributor.
	Faulty fuel pump giving top end fuel starvation	Remove, overhaul, or fit exchange reconditioned fuel pump.
Excessive oil consumption		
Oil being burnt by engine	Badly worn, perished or missing valve stem oil seals	Remove, fit new oil seals to valve stems.
	Excessively worn valve stems and valve guides	Remove cylinder head and fit new valves and valve guides.
	Worn piston rings	Fit oil control rings to existing pistons or purchase new pistons.
	Worn pistons and cylinder bores	Fit new pistons and rings, rebore cylinders.
	Excessive piston ring gap allowing blow-by	Fit new piston rings and set gap correctly.
	Piston oil return holes choked	Decarbonise engine and pistons.
Oil being lost due to leaks	Leaking oil filter gasket	Inspect and fit new gasket as necessary.
	Leaking timing case gasket	Inspect and fit new gasket as necessary.
	Leaking sump gasket	Inspect and fit new gasket as necessary.
	Loose sump plug	Tighten, fit new gasket if necessary.
Unusual noises from engine		
Excessive clearances due to mechanical wear	Worn valve gear (noisy tapping from rocker box)	Inspect and renew rocker shaft, rocker arms, and ball pins as necessary.
	Worn big-end bearings (regular heavy knocking)	Drop sump, if bearings broken up clean out oil pump and oilways, fit new bearings. If bearings not broken but worn fit bearing shells.
	Worn timing chain and gears (rattling from front of engine)	Remove timing cover, fit new timing wheels and timing chain.
	Worn main bearings (rumbling and vibration)	Drop sump, remove crankshaft; if bearings worn but not broken up, renew. If broken up strip oil pump and clean out oilways.
	Worn crankshaft (knocking, rumbling and vibration)	Regrind crankshaft, fit new main and big-end bearings.

Chapter 2 Cooling system

Contents

Antifreeze solution ... 13	General description ... 1
Cooling system - draining ... 2	Radiator - removal, inspection, cleaning and refitting ... 5
Cooling system - filling ... 4	Temperature gauge and sender unit - removing and refitting ... 12
Cooling system - flushing ... 3	Temperature gauge - fault finding ... 11
Fan belt - adjustment ... 9	Thermostat - removal, testing and refitting ... 6
Fan belt - removal and refitting ... 10	Water pump - dismantling and reassembly ... 8
Fault diagnosis - cooling system ... 14	Water pump - removal and refitting ... 7

Specifications

Type of system
Type ... Pressurised, forced circulation
Type ... Ford long life cooling fluid
Coolant mixture ... 45%

Coolant capacity
With heater ... 4.65 litres (8.19 Imp. pints)

Radiator
Type ... Tube + fin
Radiator cap type ... Bayonet catch
Pressure relief valve opens ... 0.9 to 1.1 kg cm^2 (13 lb in^2)

Thermostat
Type ... Wax capsule
Opening commences ... 85 to 98°C (185 to 192°F)
Fully open ... 99 to 102°C (210 to 216°F)

Water pump
Type ... Centrifugal
Drive ... 'V' belt

Fan belt free play ... 0.5 in (13 mm) at middle of longest span

Torque wrench settings
	lb f ft	kg fm
Fan bolts ...	5 to 7	0.69 to 0.97
Water pump ...	5 to 7	0.69 to 0.97
Thermostat housing ...	12 to 15	1.66 to 2.07

1 General description

The engine cooling water is circulated by a thermo-syphon, water pump assisted, system, and the whole system is pressurised. This is to prevent the loss of water down the overflow pipe with the radiator cap in position and to prevent premature boiling in adverse conditions. The radiator cap is pressurised to 13 lb/in^2. This has the effect of considerably increasing the boiling point of the coolant. If the water temperature goes above this increased boiling point the extra pressure in the system forces the internal part of the cap off its seat, thus exposing the overflow pipe down which the steam from the boiling water escapes, thereby relieving the pressure. It is therefore important, to check that the radiator cap is in good condition and that the spring behind the sealing washer has not weakened. Most garages have a special machine in which radiator caps can be tested. The cooling system comprises the radiator, top and bottom water hoses, heater hoses, the impeller water pump (mounted on the front of the engine, it carries the fan blades, and is driven by the fan belt), the thermostat and the cylinder block drain plug. The inlet manifold is water heated and also the automatic choke.

The system functions in the following fashion. Cold water in the bottom of the radiator circulates up the lower radiator hose to the water pump where it is pumped round the water passages in the cylinder block, helping to keep the cylinder bores and pistons cool.

The water then travels up into the cylinder head and circulates round the combustion spaces and valve seats absorbing more heat, and then, when the engine is at its proper operating temperature, travels out of the cylinder head, past the open thermostat into the upper radiator hose and into the radiator header tank.

The water travels down the radiator where it is rapidly cooled by the in-rush of cold air through the radiator core, which is created by both the fan and the motion of the car. The water, now cold, reaches the bottom of the radiator, and the cycle is repeated.

When the engine is cold the thermostat (a valve which opens and closes according to the temperature of the water) maintains the circulation of the same water in the engine.

Only when the correct minimum operating temperature has been reached, as shown in the Specifications, does the thermostat begin to open, allowing water to return to the radiator.

Chapter 2/Cooling system

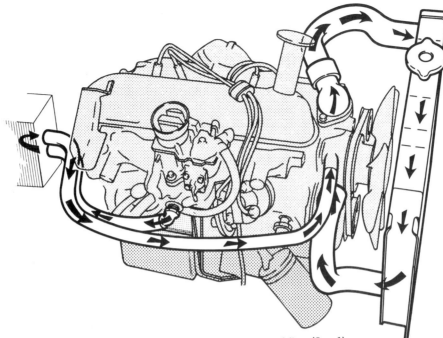

Fig. 2.1. The cooling system direction of flow (Sec. 1)

2 Cooling system - draining

1 With the car on level ground drain the system as follows:
2 If the engine is cold remove the filler cap from the radiator by turning the cap anticlockwise. If the engine is hot, having just been run, then turn the filler cap very slightly until the pressure in the system has had time to disperse. Use a rag over the cap to protect your hand from escaping steam. If, with the engine very hot, the cap is released suddenly, the drop in pressure can result in the water boiling. With the pressure released the cap can be removed.
3 If antifreeze is in the radiator, drain it into a clean bucket or bowl for re-use.
4 Remove the cylinder block drain plug, located on the rear left-hand side of the block, using a suitably sized spanner.
5 As there is no radiator drain plug fitted to the radiator the system must be further drained by slackening the radiator bottom hose clip and pulling the hose off the radiator outlet.

3 Cooling system - flushing

1 Provided the coolant is kept to its recommended concentration with antifreeze and it is renewed at the recommended intervals, flushing will not usually be required. However, due to neglect or oil entering the system because of a faulty gasket the radiator may become choked with rust scales, deposits from the water and other sediment.
2 To flush the radiator it must first be removed, as described in Section 5.
3 Ensure that the radiator cap is in place, then turn the radiator upside down and, using a high pressure supply, insert a hose in the lower hose stub and force water through the radiator and out of the upper hose stub.
4 Similarly insert the hose in the thermostat housing (after first removing the thermostat) and force water through the engine and out of the lower hose (Fig. 2.2).
5 Continue flushing both radiator and engine until the emerging water runs clean.
6 Refit the radiator, thermostat housing and hose connections.

4 Cooling system - filling

1 Ensure that the cylinder block drain plug is securely tightened and

Fig. 2.2. Back flushing the engine using a high pressure hose (arrowed) in the thermostat outlet (Sec. 2)

that the radiator bottom hose is connected.
2 Fill the system slowly to ensure that no airlocks develop. If a heater is fitted, check that the valve to the heater unit (where fitted) is open, otherwise an airlock may form in the heater. The best type of water to use in the cooling system is rainwater, so use this whenever possible.
3 Do not fill the system higher than within 0.5 inch (13 mm) of the filler orifice. Overfilling will merely result in wastage, which is to be avoided when antifreeze is in use.
4 Only use antifreeze mixture with a glycol or ethylene base (Section 13).
5 Replace the filler cap and turn it firmly clockwise to lock it in position.

5 Radiator - removal, inspection, cleaning and refitting

1 To remove the radiator first drain the cooling system, as described in Section 2.
2 Slacken the clip securing the radiator top hose to the radiator and pull the hose off the radiator stub.
3 Undo the five bolts and remove the air deflector panel (Fig. 2.3) between the radiator and front panel.
4 At the rear of the radiator there is a cowling, designed to channel

the airflow through the vehicle onto the fan and engine. This cowling is secured to the radiator with four bolts and washers. These must now be removed and the cowling detached and placed to one side (over the fan blades) (Figs. 2.4 and 2.5).

5 Undo and remove the two bolts and washers on each side of the radiator which hold it in place then lift the radiator out of the engine compartment.

6 With the radiator out of the car any leaks can be soldered or repaired. Clean out the inside of the radiator by flushing as detailed in Section 3. When the radiator is out of the car, it is advantageous to turn it upside down for reverse flushing. Clean the exterior of the radiator by hosing down the radiator matrix with a strong jet of water to clear away road dirt, dead flies etc.

7 Inspect the radiator hoses for cracks, internal or external perishing, and damage caused by overtightening of the securing clips. Replace the hoses as necessary. Examine the radiator hose securing clips and renew them if they are rusted or distorted.

8 Refitting is a straightforward reversal of the removal procedure.

6 Thermostat - removal, testing and refitting

1 To remove the thermostat, partially drain the cooling system (four pints is enough), then loosen the wire clip retaining the top radiator hose to the outlet elbow and pull the hose off the elbow.

2 Undo the two bolts holding the elbow to the cylinder head and remove the elbow and gasket.

3 The thermostat can now be lifted out. Should the thermostat be stuck in its seat, do not lever it upwards but cut through the corrosion all round the edge of the seat with a sharp pointed knife. This will usually release the thermostat without causing any damage (Fig. 2.6).

4 Test the thermostat for correct functioning by suspending it by a length of string in a saucepan of cold water together with a thermometer (Fig. 2.7).

5 Heat the water and note when the thermostat begins to open. The correct opening temperature is stamped on the flange of the thermostat and is also given in the Specifications.

6 Discard the thermostat if it opens too early. Continue heating the water until the thermostat is fully open. Then let it cool down naturally. If the thermostat will not open fully in boiling water, or does not close down as the water cools, then it must be renewed.

7 If the thermostat is stuck open when cold this will be apparent when removing it from the housing.

8 Replacing the thermostat is a reversal of the removal procedure. Remember to use a new gasket between the elbow and the cylinder head. If any pitting or corrosion is apparent, it is advisable to apply a layer of sealing compound such as Golden Hermetite to the metal surfaces of the housing before reassembly. If the elbow is badly eaten away it must be replaced with a new component.

Fig. 2.3. Remove the air deflection panel (Sec. 5)

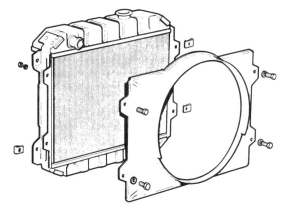

Fig. 2.4. Radiator cowl fixing to radiator (Sec. 5)

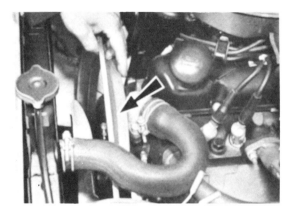

Fig. 2.5. Hanging the radiator cowl (arrowed) over the fan blades (Sec. 5)

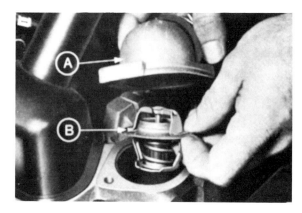

Fig. 2.6. Lifting the thermostat housing and top hose (A) from the cylinder head and removing the thermostat (B) (Sec. 6)

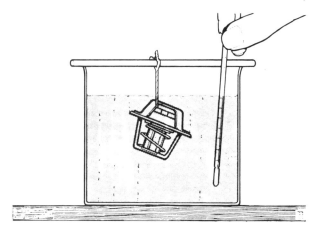

Fig. 2.7. Testing the thermostat (Sec. 6)

Chapter 2/Cooling system

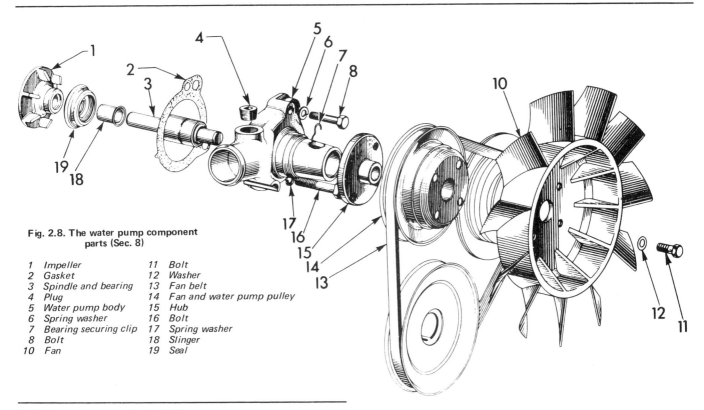

Fig. 2.8. The water pump component parts (Sec. 8)

1　Impeller
2　Gasket
3　Spindle and bearing
4　Plug
5　Water pump body
6　Spring washer
7　Bearing securing clip
8　Bolt
10　Fan
11　Bolt
12　Washer
13　Fan belt
14　Fan and water pump pulley
15　Hub
16　Bolt
17　Spring washer
18　Slinger
19　Seal

7　Water pump - removal and refitting

1　Drain the cooling system, as described in Section 2, then undo the clip on the small heater hose and pull the hose off the pump.
2　Undo and remove the alternator adjustment arm bolt, slacken the two bolts securing the alternator to the mounting bracket and push the alternator towards the engine. Prise the fan belt from the alternator and water pump pulleys and then from around the crankshaft pulley. (Fig. 2.9).
3　Remove the radiator, as described in Section 5.
4　Undo the four bolts and washers which hold the fan and the pulley wheel in place (Fig. 2.10).
5　Remove the fan and the pulley wheel and then undo the three bolts holding the water pump in place and withdraw the pump together with its gasket (Fig. 2.11).
6　Replacement is a reversal of the above procedure but always remember to use a new gasket. When adjusting the fan belt tension ensure that there is 0.5 in (13 mm) total movement at the centre of the span between the alternator and water pump pulleys before finally tightening the alternator mounting bolts (Fig. 2.12)..

8　Water pump - dismantling and reassembly

Note: All numbers used in this Section refer to Fig. 2.8.
1　Remove the hub (15) from the water pump shaft (3) by using a suitable hub puller.
2　Carefully pull out the bearing retainer wire (7) and then with the aid of two blocks (a small mandrel and a large vice, if the proper tools are not available) press out the shaft and bearing assembly (3) together with the impeller (1) and seal from the water pump body (5).
3　The impeller vane is removed from the spindle with an extractor.
4　Remove the seal (19) and the slinger (18) by splitting the latter with the aid of a sharp cold chisel.
5　The repair kit available comprises a new shaft and bearing assembly, a slinger, seal, bush, clip and gasket.
6　To reassemble the water pump, press the shaft and bearing assembly (3) into the housing with the short end of the shaft to the front, until the groove in the shaft is in line with the groove in the housing. The bearing retainer wire (7) can then be inserted.
7　Press the pulley hub (15) onto the front end of the shaft (3) until the end of the shaft is half an inch from the outer face of the hub.
8　Fit the new slinger bush (18) with the flanged end first onto the

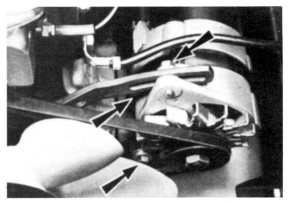

Fig. 2.9. Alternator mounting bolt location. These bolts (arrowed) must be slackened before the alternator can be moved (Sec. 7)

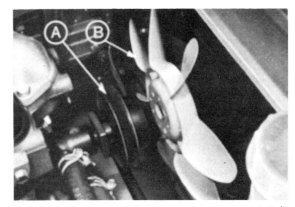

Fig. 2.10. Fan (B) and pulley (A) assembly on water pump hub (Sec. 7)

Fig. 2.11. Removing the water pump (Sec. 7)

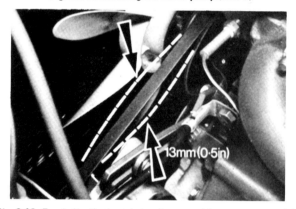

Fig. 2.12. Fan belt adjustment showing a total free-movement of 13 mm (0.5 in) at the mid-point of the longest span of the belt (Sec. 7)

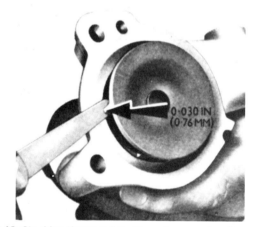

Fig. 2.13. Checking clearance between water pump impeller and body (Sec. 8)

rear of the shaft (3) and refit the pump seal (19) with the thrust face towards the impeller (1).
9 Press the impeller (1) onto the shaft (3) until a clearance of 0.030 in (0.76 mm) is obtained between the impeller blades and the housing face as shown in Fig. 2.13.
10 It is important to check at this stage that the pump turns freely and smoothly before replacement onto the block. After replacement check carefully for leaks.

9 Fan belt - adjustment

1 The fan belt tension is correct when there is 0.5 in (13 mm) of lateral movement at the midpoint position of the belt between the

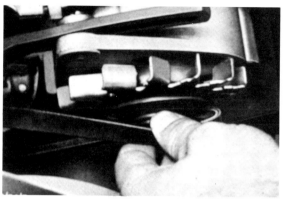

Fig. 2.14. Removing the belt from alternator pulley (Sec. 10)

alternator pulley wheel and the water pump pulley wheel.
2 To adjust the fan belt, slacken the alternator securing bolts and move the alternator either in or out until the correct tension is obtained. It is easier if the alternator securing bolts are only slackened slightly so it requires some force to move the alternator. In this way the tension of the belt can be arrived at more quickly than by making frequent adjustments.
3 If difficulty is experienced in moving the alternator away from the engine, a long spanner or screwdriver placed behind the alternator and resting against the cylinder block serves as a very good lever and can be held in this position while the alternator securing bolts are tightened down.

10 Fan belt - removal and refitting

1 If the fan belt is worn or has stretched unduly it should be renewed. The most usual reason for renewal is that the belt has broken in service. It is therefore recommended that a spare belt is always carried.
2 To remove the belt loosen the alternator securing bolts and push the alternator in towards the engine.
3 Slip the old belt over the crankshaft, alternator and water pump pulley wheels and lift it off over the fan blades. (Fig. 2.14).
4 Put on a new belt in the same way and adjust it as described in the previous Section. Note: after fitting a new belt it will require adjustment due to its initial stretch after about 250 miles (400 km).

11 Temperature gauge - fault finding

1 If the temperature gauge fails to work, either the gauge, the sender unit, the wiring or the connections are at fault.
2 It is not possible to repair the gauge or the sender unit and they must be replaced by new units if at fault.
3 First check the wiring for breaks using an ohmmeter or continuity tester. The sender unit and gauge should be tested by substitution.

12 Temperature gauge and sender unit - removal and refitting

1 For details of how to remove and replace the temperature gauge see Chapter 10.
2 To remove the sender unit, drain half the coolant from the system, disconnect the wire leading into the unit at its connector and undo the unit with a spanner. The unit is located in the cylinder head just below the water outlet elbow on the left side. Replacement is a reversal of the above procedure.

13 Antifreeze solution

1 Apart from the protection against freezing conditions which the use of antifreeze provides, it is essential to minimise corrosion in the cooling system.
2 The cooling system is initially filled with a solution of 45%

antifreeze and it is recommended that this percentage is maintained.
3 With long-life types of antifreeze mixtures, renew the coolant every two years. With other types, drain and refill the system every twelve months.
4 The following table gives a guide to protection against frost but a mixture of less than 30% concentration will not give protection against corrosion:

Amount of antifreeze	Protection to
45%	−32°C (−26°F)
40%	−25°C (−13°F)
30%	−16°C (+ 3°F)
25%	−13°C (+ 9°F)
20%	− 9°C (+15°F)
15%	− 7°C (+20°F)

14 Fault diagnosis - cooling system

Symptom	Reason/s	Remedy
Overheating Heat generated in cylinder not being successfully disposed of by radiator	Insufficient water in cooling system	Top up radiator.
	Fan belt slipping (accompanied by a shrieking noise on rapid engine acceleration)	Tighten fan belt to recommended tension or replace if worn.
	Radiator core blocked or radiator grille restricted	Reverse flush radiator, remove obstructions.
	Bottom water hose collapsed, impeding flow	Remove and fit new hose.
	Thermostat not opening properly	Remove and fit new thermostat.
	Ignition advance and retard incorrectly set (accompanied by loss of power, and perhaps, misfiring)	Check and reset ignition timing.
	Carburettor(s) incorrectly adjusted (mixture too weak)	Tune carburettor(s).
	Exhaust system partially blocked	Check exhaust pipe for constrictive dents and blockages.
	Oil level in sump too low	Top up sump to full mark on dipstick.
	Blown cylinder head gasket (water/steam being forced down the radiator overflow pipe under pressure)	Remove cylinder head, fit new gasket.
	Engine not yet run-in	Run-in slowly and carefully.
	Brakes binding	Check and adjust brakes if necessary.
Engine runs cool Too much heat being dispersed by radiator	Thermostat jammed open	Remove and renew thermostat.
	Incorrect grade of thermostat fitted allowing premature opening of valve	Remove and replace with new thermostat which opens at a higher temperature.
	Thermostat missing	Check and fit correct thermostat.
Loss of cooling water Leaks in system	Loose clips on water hoses	Check and tighten clips if necessary.
	Top, bottom, or by-pass water hoses perished and leaking	Check and replace any faulty hoses.
	Radiator core leaking	Remove radiator and repair.
	Thermostat gasket leaking	Inspect and renew gasket.
	Radiator pressure cap spring worn or seal ineffective	Renew radiator pressure cap.
	Blown cylinder head gasket (pressure in system forcing water/steam down overflow pipe)	Remove cylinder head and fit new gasket.
	Cylinder wall or head cracked	Dismantle engine, dispatch to engineering works for repair.

Chapter 3 Carburation; fuel and exhaust systems

For modifications, and information applicable to later models, see Supplement at end of manual

Contents

Accelerator cable - removal, refitting and adjustment	18	Fuel tank - removal and refitting	6
Accelerator pedal and pedal shaft - removal and refitting	19	General description	1
Air cleaner - removal, servicing and refitting	2	Motorcraft single venturi carburettor - accelerator pump adjustment	15
Carburettor - Motorcraft single venturi - general description	10	Motorcraft single venturi carburettor - choke plate pull down adjustment	14
Exhaust pipe unit - removal and refitting	21		
Exhaust system - general description	20		
Fault diagnosis - carburation; fuel and exhaust systems	24	Motorcraft single venturi carburettor - cleaning	12
Fuel gauge sender unit - removal and refitting	8	Motorcraft single venturi carburettor - float setting	13
Fuel pump - removal and refitting	3	Motorcraft single venturi carburettor - removal and refitting	11
Fuel pump - servicing	5	Motorcraft single venturi carburettor - fast idling adjustment	17
Fuel pump - testing	4	Motorcraft single venturi carburettor - slow running adjustment	16
Fuel tank - cleaning and repair	7	Muffler unit - removal and refitting	22
Fuel tank filler pipe - removal and refitting	9	Resonator unit - removal and refitting	23

Specifications

Air filter
Element material	Paper
Type	Dry

Fuel line
Material	Cotton braided PVC nitrile
Inner diameter (mm)	7.47 ± 0.38

Fuel tank
Material finish	Coated steel
Mounting	Retaining straps
Capacity	12.7 gallons (58 litres)

Fuel pump
Type	Diaphragm
Drive	Mechanical from camshaft
Location	Cylinder block, right-hand side rear
Delivery pressure	3.0 to 5.0 lb in^2 (0.21 to 0.35 kg/cm^2)

Carburettor
Type	Motorcraft		
	761 F - 9510 - AA	71 HF - 9510 - KDA	761 F - 9510 - KBA
Choke plate pull down setting	3 mm (0.12 in)	2.25 ± 0.25 mm (0.09 ± 0.01 in)	3.0 ± 0.25 mm (0.12 ± 0.01 in)
Throttle barrel diameter	32 mm (1.25 in)	36 mm (1.4 in)	34 mm (1.3 in)
Venturi diameter	23/24 mm (0.90/0.94 in)	28 mm (1.1 in)	25 mm (0.98 in)
Main jet	120/105 mm (4.72/4.0 in)	137 mm (5.4 in)	122 mm (4.7 in)
Idling speed (rpm)	800 ± 25	750 ± 25	800 ± 25
Fast idle (rpm)	—	2000 ± 100	1400 ± 100
Float level	41.00 ± 0.3 mm (1.61 ± 0.01 in)	27.9 ± 0.75 mm (1.10 ± 0.03 in)	29.0 ± 0.75 mm (1.14 ± 0.03 in)
Float travel	11.5 mm (0.45 in)	7.10 mm (0.28 in)	—
Idle mixture % CO	—	—	1.6 ± 0.2

Fuel octane requirement	97 octane

Torque wrench settings
	lb f ft	kg fm
Air cleaner to carburettor or rocker cover	4 to 7	0.6 to 0.9
Air cleaner lid to body	4 to 5	0.5 to 0.7
Carburettor to inlet manifold	12 to 15	1.7 to 2.1
Fuel pump to engine	12 to 15	1.7 to 2.1

Chapter 3/Carburation; fuel and exhaust systems

1 General description

The fuel system comprises a 12.7 gallon (58 litre) fuel tank, a mechanically operated fuel pump and a Motorcraft single venturi carburettor.

The fuel tank is located in the luggage compartment below the floor, and is retained by two straps. The fuel outlet and sender unit are combined in the front face of the tank. A breather pipe is located in the right-hand load space trim panel.

The diaphragm type fuel pump is located on the right-hand side of the cylinder block and is operated by the camshaft.

The fuel pump has a nylon screen filter which can be removed for cleaning.

2 Air cleaner - removal, servicing and refitting

1 The renewable paper element air cleaner is fitted onto the top of the carburettor installation. Servicing is confined to cleaning, or renewal, of the element at the specified service intervals.
2 Every 6000 miles (96000 km) tap the element on a hard surface or use compressed air from a tyre pump to remove surface dust. Never attempt to clean it in solvent or petrol.
3 Every 18,000 miles (29,000 km) renew the element. Always check the condition of the rubber sealing rings and renew them if they are perished or deformed.
4 To remove the element, remove the three screws from the top of the air cleaner unit and lift clear the lid.
5 Withdraw the element.
6 Should the air cleaner unit require to be removed, for example if the carburettor is to be serviced, this is easily done by simply pressing the three air cleaner stays off the lugs on the air cleaner body.
7 Refitting is the reversal of removing but be sure to seat the lid correctly before tightening the self-tapping screws.

3 Fuel pump - removal and refitting

1 Disconnect the fuel inlet and outlet pipes. Where crimped hose clips are fitted these should be discarded and replaced by a screw type hose clamp.
2 Unscrew and remove the two securing bolts from the pump flange and remove the pump with gasket from the crankcase.
3 Refitting is a reversal of removal but use a new gasket and make sure that the pump rocker arm is correctly positioned on top of the camshaft eccentric. Tighten the bolts to a torque of between 12 - 15 lb f ft (1.7 - 2.1 kg fm).

4 Fuel pump - testing

1 To test the pump fitted in position on the crankcase, detach the fuel inlet pipe at the carburettor and disconnect the HT lead from the ignition coil.
2 Operate the starter switch which should cause well defined spurts of petrol to be ejected from the disconnected end of the pipe.
3 If the pump is removed from the engine, place a finger over the inlet port and work the rocker arm several times. Remove the finger - a distinct suction noise should be heard.
4 Now place a finger over the outlet port, depress the rocker arm to its fullest extent and immerse the pump in paraffin. Watch for air bubbles which would indicate leakage at the pump flanges.

5 Fuel pump - servicing

1 Remove the single screw securing the fuel pump cover to the pump body and lift off the cover, then detach the cover seal and withdraw the 'top hat' filter from the pump body (Fig. 3.3).
2 Thoroughly clean the cover, filter and pump body, using a paintbrush and clean petrol to remove any sediment.
3 Reassemble the pump and carry out the test detailed in Section 4. Should the pump prove to be in need of attention it will have to be renewed as a complete unit, as it is not possible to dismantle it, or obtain spare parts.

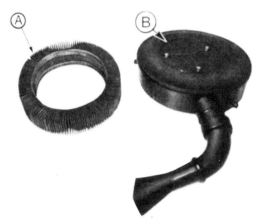

Fig. 3.1. The element 'A' and air cleaner unit 'B' (Sec. 2)

Fig. 3.2. Remove existing fuel pump clip and replace with a screw type hose clip (Sec. 3)

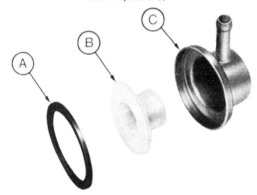

Fig. 3.3. Fuel pump servicing: (A) Seal (B) Filter (C) Cover (Sec. 5)

6 Fuel tank - removal and refitting

1 Disconnect the battery earth lead.
2 Syphon any fuel remaining in the tank into a clean petrol or oil can. Seal the can and store in a safe place.
3 Jack up the rear of the car and fit axle stands or blocks to support it. Place chocks under the front wheels.
4 From the front of the tank disconnect the fuel feed pipe from its connection, Fig. 3.4, and unclip it from the retaining clips along the front edge.
5 Pull clear the two wires from the sender unit.
6 From the chassis unclip the vent pipe and undo the breather pipe at the 'T' connection.
7 Slacken the securing straps (Fig. 3.5) whilst supporting the tank. With the straps unclipped, the tank unit can be removed complete with guard if fitted. Note that the filler pipe remains in position.
8 Refitting is a reversal of the above procedure but ensure that the

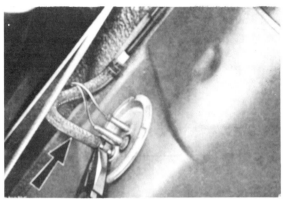

Fig. 3.4. Disconnect the fuel feed pipe (Sec. 6)

Fig. 3.5. The fuel tank assembly (Sec. 6)

A Seal
B Rubber insulators
C Sender unit
D Securing straps
E 'T' connection

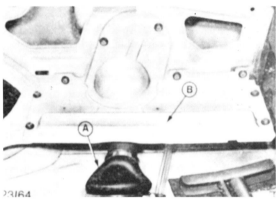

Fig. 3.6. Fuel tank filler pipe location (Sec. 9)

A Filler pipe gaiter
B Cover panel

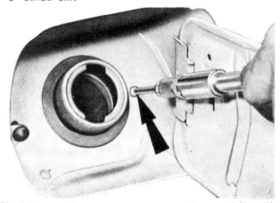

Fig. 3.7. Upper filler neck retaining screw (arrowed) (Sec. 9)

rubber insulating pads are correctly positioned and stuck to the tank.
  Be sure to fit the filler pipe seal and, to ease assembly, lightly grease the filler tank neck, and check when the tank is in position that the filler pipe is fully located within the tank.
  Finally, when the tank is fully fitted and refilled with fuel, check for leaks.

7 Fuel tank - cleaning and repair

1 With time it is likely that sediment will collect in the bottom of the fuel tank. Condensation, resulting in rust and other impurities, will usually be found in the fuel tank of any car more than three or four years old.
2 When the tank is removed, it should be swilled out using several changes of paraffin and finally rinsed out with clean petrol. Remember that the float mechanism is delicate and the tank should not be shaken violently or turned upside down quickly in case damage to the sender unit is incurred.
3 If the tank is leaking it should be renewed or taken to a specialist firm for repair. Do not attempt to solder, braze or weld it yourself, it can be lethal. A temporary repair may be made with fibreglass or similar material but a new tank should be fitted as quickly as possible.

8 Fuel gauge sender unit - removal and refitting

1 The sender unit can be removed with the tank in the car. Follow the instructions given in Section 6, paragraphs 1 to 5.
2 Unscrew (anticlockwise) the sender unit retaining ring using a suitable 'C' spanner or by tapping the projections carefully with a hammer and cold chisel. Remove the sealing ring.
3 Refitting is a reversal of the removal procedure but always fit a new sealing ring and refit the sender unit carefully - it is a delicate component.

9 Fuel tank filler pipe - removal and refitting

1 Remove the fuel tank as described in Section 6.
2 Remove the right-hand rear trim panel in the loading space.
3 Withdraw the spare wheel covering panel.
4 Remove the internal filler pipe panel by unscrewing the nine retaining bolts (Fig. 3.6).
5 Slacken the floor to pipe gaiter clamps and withdraw the gaiter from the floor panel.
6 Now remove the filler cap and unscrew the retaining screw from the body to filler neck (Fig. 3.7).
7 The filler pipe can now be detached and the gaiter removed.
8 Refitting is the reverse procedure of the above, but check for leaks when refitting the tank before refitting the trim panel.

10 Carburettor - general description

Prior to build code RL (mid-May 1975)
  This carburettor is of the single venturi downdraught design incorporating an accelerator pump and power valve, as well as the usual engine idle and main systems.
  The carburettor body comprises two parts: ie. the upper and lower bodies, with an additional two piece cast housing for the auto choke system.
  The upper body forms the cover to the float chamber and houses the float pivot brackets, float, fuel inlet connection tube, main jet, float retaining pin and choke plate.
  The lower body houses the following major components: throttle barrel with integral choke tube, throttle and lever, adjustment screws, accelerator pump and the distributor auto-advance connection.

Chapter 3/Carburation; fuel and exhaust systems

Build code RL (mid-May 1975) onwards

This carburettor differs from the type previously used in that it incorporates the 'By-pass' (Sonic) idle system which was introduced for improved economy. Where installed, it can be instantly recognised by the seven securing screws for the upper body compared with six for the previous carburettor, and the increased length of the vacuum pick-up pipe (35 mm/1.37 in compared with 14 mm/0.52 in).

The Sonic idle system differs from the conventional idle system in that the majority of the idle air flow, and all of the idle fuel flow passes through the by-pass system. The remainder of the idle air flow passes through the gap provided by the carburettor butterfly being held fractionally open in the idle condition.

Build code SC (May 1976) onwards

This carburettor differs from the build code RL type in that the idle mixture adjustment screw is sealed by a white plastic plug to prevent unauthorized adjustment. The carburettor is designed so that after the initial running-in period of a new engine, the correct exhaust gas CO content will be obtained; this should remain at an acceptable level (ie. no idle mixture adjustment required) throughout the service life of the carburettor.

Where idle mixture adjustment is found to be necessary, the white plastic plug can be punctured in its centre using a small diameter stiff wire, and the plug prised out. The idle mixture may then be adjusted in the normal way (see Section 16), but a replacement plug must be pressed in on completion.

Note that during any adjustment of these carburettors, a proprietary exhaust gas analyzer must be used to check that the CO content of the exhaust gas is within the specified limits (also see Section 16).

11 Motorcraft single venturi carburettor - removal and refitting

1 Remove the air cleaner and disconnect the vacuum and fuel inlet pipes from the carburettor.
2 Free the throttle shaft from the throttle lever by sliding back the securing clip and undo the screw which holds the end of the choke cable in place.
3 Undo the two nuts and spring washers which hold the carburettor in place and lift the carburettor off the inlet manifold.
4 Replacement is a straightforward reversal of the removal sequence but note the following points:
   a) *Remove the old inlet manifold to carburettor gasket, clean the mating flanges and fit a new gasket in place. Where a Sonic idle carburettor is used, or where an inlet manifold has been renewed, ensure that the carburettor/manifold gasket tab is towards the front of the engine.*
   b) *If the fuel line hose is retained by a crimped type clamp this must be replaced with a screw type.*
   c) *Ensure that the choke knob is in the off position before connecting the inner choke cable at the carburettor. After connection ensure that the choke opens and closes fully with a very slight amount of slack in the cable when the choke control is pushed right in.*

12 Motorcraft single venturi carburettor - cleaning

1 Thoroughly clean the exterior surfaces of the carburettor body.
2 Undo the six screws and washers which hold the carburettor top to the main body.
3 Lift off the top from the main body (Fig. 3.11), at the same time

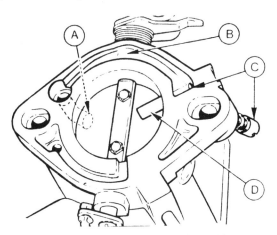

Fig. 3.8. By-pass (Sonic) idle system (Sec. 10)

A Air intake for by-pass system
B Air distribution channel
C Mixture (volume control) adjustment screw
D Sonic discharge tube

Fig. 3.9. Arrows indicate items to be disconnected to remove the carburettor (Sec. 11)

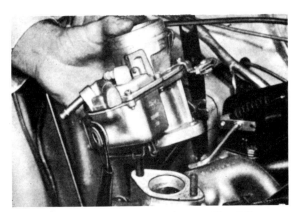

Fig. 3.10. Removing the carburettor (Sec. 11)

Fig. 3.11. Removing the carburettor upper body (Sec. 12)

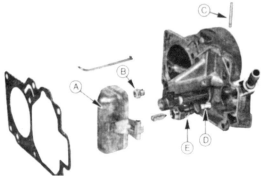

Fig. 3.12. The component parts of the upper body: (Sec. 12)

A Float
B Main jet
C Float retaining pin
D Intake filler
E Valve housing

Fig. 3.13. Carefully remove the accelerator pump assembly: (Sec. 12)

A Check valve spring
B Diaphragm return spring

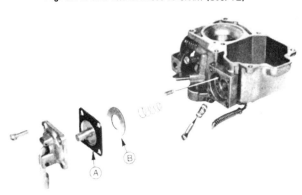

Fig. 3.14. Jets and orifices to clean (Sec. 12)

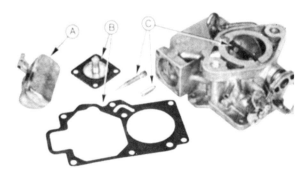

Fig. 3.15. Components to check for wear/damage (Sec. 12)

Fig. 3.16. The accelerator pump assembly: (Sec. 12)

A Diaphragm
B Sealing washer

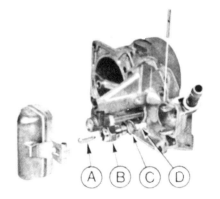

Fig. 3.17. Float assembly (Sec. 12)

A Needle
B Needle valve housing
C Seal
D Filter

Fig. 3.18. Accelerator pump discharge ball valve (B) and weight (A) (Sec. 12)

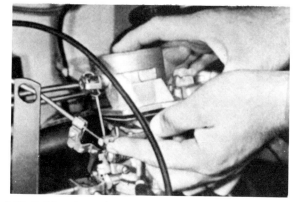

Fig. 3.19. Holding the choke mechanism closed while refitting the upper body (Sec. 12)

Chapter 3/Carburation; fuel and exhaust systems

unlatching the choke control rod. Ensure that the gasket comes off with the top cover.

4 Turn the carburettor lower body upside-down and allow the accelerator pump discharge ball valve and weight to fall out.

5 From the top cover pull out the float retaining pin and detach the float. With the float removed the needle valve housing assembly can be unscrewed and the upper body gasket lifted off the upper body. Finally, remove the needle valve housing seal and filter, and the main jet from their location in the underside of the upper body.

6 Remove the four screws securing the accelerator pump assembly to the side of the lower body and withdraw the pump from the body, taking care not to lose the two return springs.

7 Remove the volume control (mixture) screw from its location in the lower body. (Refer to Section 10 if a sealed mixture screw is installed).

8 Clean the carburettor and float chamber jets, paying particular attention to those locations shown in Fig. 3.14. Then inspect the float for petrol penetration (ie. holes in the float wall), the pump diaphragm and gasket for splits, and the mixture screw and needle valve for distortion and wear.

9 Refit the volume control screw.

10 Reassemble the accelerator pump assembly in the lower housing, with the steel side of the sealing washer facing away from the body and the tapered end of the spring toward the body.

11 Refit the needle valve, float and main jet in the reverse sequence to their removal, making sure that the needle valve seal and filter are fitted to the needle valve housing **before** the housing is screwed into the upper body.

12 Check and adjust the float level as described in Section 13.

13 Insert the accelerator pump discharge ball valve and weight into their location in the lower body. Align the upper body gasket, reconnect the choke operating linkage and then align and secure the upper body to the lower body with six screws. During this operation ensure that (a) the cranked end of the choke link is fitted at the bottom, and (b) when fitting the upper body ensure that the choke mechanism is held in the fully closed position; this will avoid the choke cam overcentering during the installation.

14 Check and adjust the choke plate pull down as described in Section 14.

15 Check and adjust the accelerator pump stroke as described in Section 15.

13 Motorcraft single venturi carburettor - float setting

1 The height of the float is important in the maintenance of a correct flow of fuel. The correct height is determined by measurement and by bending the tab which rests on the end of the needle valve. If the height of the float is incorrect there will either be fuel starvation symptoms or fuel will leak from the joint of the float chamber.

2 To check the fuel level setting turn the carburettor upper body to a vertical position so that the float closes the needle valve by its own weight. This corresponds to its true position in the float chamber when the needle valve is closed and no more fuel can enter the chamber. Remove the upper body gasket before checking.

3 Measure the distance from the normal base of the float and the metal face of the underside of the upper body. This should be 1.16 in ± 0.01 in (29.5 ± 0.25 mm). If this measurement is not correct then bend the tab which rests on the needle valve until the correct measurement is obtained.

4 Turn the body the right way up and take the same measurement with the float in the fully open position. The measurement should now be 1.38 in (35 mm ± 0.25 mm). If this measurement is not correct bend the other (hinge) tab until the correct measurement is obtained.

14 Motorcraft single venturi carburettor - choke plate pull down - adjustment

1 Remove the air cleaner and rotate the choke lever until it is against its stop.

2 Depress the choke plate and check the gap between the edge of the plate and the side of the carburettor air intake as shown in Fig. 3.22. The gap is correct when it measures 3.30 mm using the shank of a drill of the correct size as a measuring instrument.

3 If the gap is incorrect bend the tab on the choke spindle until the drill will just fit.

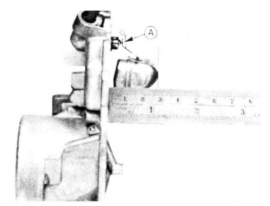

Fig. 3.20. Measuring float level: bend tab 'A' to adjust (Sec. 13)

Fig. 3.21. Measuring float travel: bend hinge 'A' to adjust (Sec. 13)

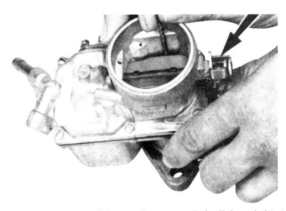

Fig. 3.22. Choke plate pull down adjustment: choke linkage held closed while gap is checked with shank of twist drill (Sec. 14)

15 Motorcraft single venturi carburettor - accelerator pump adjustment

1 Under normal conditions the accelerator pump requires no adjustment. If it is wished to check the accelerator pump action, first slacken the throttle stop screw so that the throttle plate is completely closed.

2 Press in the diaphragm plunger fully and check that there is a 2.67 mm (0.11 in) clearance between the operating lever and the plunger. The clearance is most easily checked by using a suitable sized drill (Fig. 3.23).

3 To shorten the stroke open the gooseneck of the pump pushrod, and to lengthen the stroke close the gooseneck.

4 If poor acceleration can be tolerated for maximum economy disconnect the operating lever to the accelerator pump entirely.

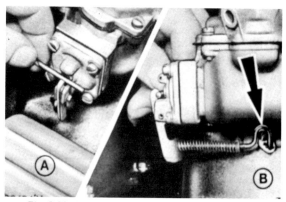

Fig. 3.23. Accelerator pump adjustment: (Sec. 15)

A Check adjustment stroke with 2.67 mm dia. drill
B Open or close gooseneck to effect adjustment

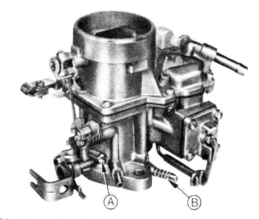

Fig. 3.24a. Carburettor adjustment screws (prior to build code RL) (Sec. 16)

A Idle speed screw
B Idle mixture screw

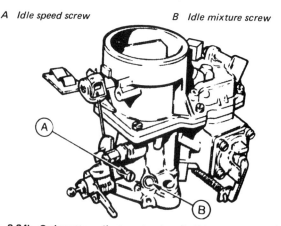

Fig. 3.24b. Carburettor adjustment points (build code SC onwards) (Secs. 10 and 16)

A Idle speed screw
B White plastic blanking plug over mixture screw

Note: On carburettors from build code RL to SC, the idle mixture screw is in the position of the white plastic blanking plug

16 Motorcraft single venturi carburettor - slow-running adjustment

Note: Refer to Section 10 if a sealed mixture screw is installed.

1 Adjustment of the carburettor should only be carried out with the engine at normal operating temperature. Tuning by ear should be regarded as a temporary expedient only and one of two recommended methods (vacuum gauge or 'Colortune') used whenever possible.
2 To adjust the slow-running by ear turn the throttle stop screw (Fig. 3.24a or 24b) so that the engine is running at a fast idle. Turn the volume (mixture) control screw in or out until the engine runs evenly without 'hunting' or 'lumpiness'. Reduce the idling speed and re-adjust the volume control screw.
3 To adjust the slow-running using a vacuum gauge, remove the blanking plug located just below the carburettor mounting flange on the inlet manifold. On vehicles fitted with a semi-closed crankcase ventilation system, remove the fume extraction hose from the manifold nozzle and substitute a tee-connector so that both the fume extraction hose and the vacuum pipe can be connected to the inlet manifold. On vehicles fitted with a semi-closed crankcase ventilation system and a brake vacuum servo unit, pull off the servo flexible hose from the inlet manifold tee-connector and substitute the vacuum gauge pipe. Set the throttle stop screw so that the engine is running at the recommended idling speed (see Specifications) and then turn the volume control screw so that the reading on the vacuum gauge is at maximum obtainable. Re-adjust both screws if necessary to reduce idling speed but maintain maximum vacuum reading.
4 To adjust the slow-running using 'Colortune', follow the manufacturer's instructions.
5 With any of these methods, satisfactory adjustment will not be obtained if there are any air leaks in the system. Check the security of the inlet manifold and carburettor flange gaskets, and particularly the rubber or plastic connectors at each end of the distributor vacuum pipe for splits or looseness.

17 Motorcraft single venturi carburettor - fast idling adjustment

1 Remove the air cleaner and rotate the choke lever until it is against its stop.
   The fast idle check and any necessary adjustment should only be made after the choke has been checked and adjusted.
2 If the engine is cold, run it until it reaches its normal operating temperature and then allow it to idle naturally.
3 Hold the choke plate in the fully open vertical position and turn the choke lever until it is stopped by the choke linkage. With the choke lever in this position, the engine speed should rise to about 1100 rev/min as the fast idle cam will have opened the throttle very slightly.
4 Check how much radial movement is needed on the throttle lever to obtain this result and then stop the engine.
5 With a pair of mole grips clamp the throttle lever fully open on the stop portion of the casting boss and bend down the tab to decrease, or up to increase, the fast idle speed.
6 Remove the grips and check again if necessary repeating the operation until the fast idling is correct. It may also be necessary to adjust the slow idling speed and recheck the choke setting.

18 Accelerator cable - removal, refitting and adjustment

1 Disconnect the battery earth lead.
2 Disconnect the accelerator inner cable at the throttle link, turn back the locknut then detach the outer cable from the bracket.
3 Detach the outer cable from the bulkhead (1 screw).
4 From inside the car, remove the dash lower insulator panel. It is retained by five screws (rhd) or three screws (lhd), along the rear edge and can then be unclipped from the front edge.
5 Remove the retaining clip from the pedal shaft by depressing at point 'A' and lifting at point 'B' (Fig. 3.25), then pull the inner cable through the shaft and lift it out of the slot.
6 Refitting is the reverse of the removal procedure, the clip on the shaft being pressed in to secure the cable end.
7 To adjust the cable, remove the air cleaner (Section 2) and detach the throttle return spring.
8 Fully slacken the outer cable adjusting nut and locknut.
9 Jam the throttle pedal in the wide open position using a block of wood or similar item.
10 Wind back the accelerator cable adjusting nut to a point where the carburettor linkage is just in the fully open position, then securely tighten the locknut.
11 Reconnect the throttle return spring then check the pedal action.
12 Refit the air cleaner (Section 2) and reconnect the battery earth lead.

Chapter 3/Carburation; fuel and exhaust systems 53

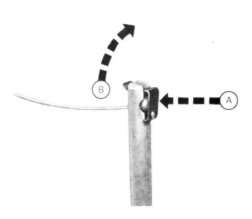

Fig. 3.25. The cable retaining clip 'A'. Unclip in direction 'B' (Sec. 18)

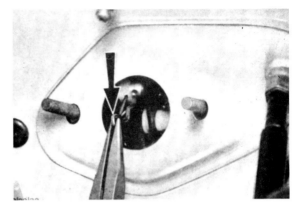

Fig. 3.26. Pull out the shaft end securing clip (Sec. 19)

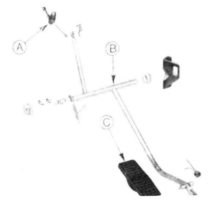

Fig. 3.27. The accelerator shaft assembly - Rhd models (Sec. 19)

A Throttle cable B Accelerator shaft C Throttle pedal

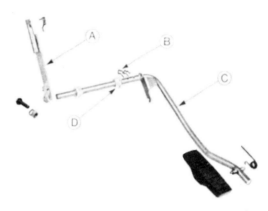

Fig. 3.28. The accelerator shaft assembly - Lhd models (Sec. 19)

A Shaft extension rod C Accelerator shaft
B Retaining spring clip D End mounting bush

19 Accelerator pedal and pedal shaft - removal and refitting

Note: If the pedal only is to be removed refer to paragraph 12.
1 Disconnect the battery earth lead.
2 From inside the car remove the dash lower insulator panel. It is retained by five screws (rhd) or three screws (lhd) along the rear edge and can then be unclipped from the front edge.
3 Remove the accelerator cable from the pedal shaft as described in the previous Section.

Rhd variants
4 Disconnect the brake operating rod at the brake pedal, then remove the master cylinder, as described in Chapter 9.
5 Working through the rear bulkhead in the engine compartment, pull out the shaft end securing clip (Fig. 3.26).
6 Rotate the right-hand shaft mounting bush through 45° in either direction and pull it out.
7 Detach the accelerator shaft assembly.

Lhd variants
8 Loosen the clamp and detach the shaft extension rod.
9 Carefully drive out the right-hand mounting bush retaining clip from the shaft then slide out the shaft until it fouls the heater box.
10 Detach the right-hand mounting bush from the pedal box by rotating it through 45° in either direction, then pulling it out. The accelerator shaft assembly can now be removed.

All models
11 Detach the remaining bush and clip from the shaft.
12 To remove the pedal, prise the flange away from the spigot on the shaft, then remove the pedal and spring.
13 When refitting the pedal, locate the spring on the spigot shaft then clip the flanges onto the spigots and check that the pedal pivots correctly.
14 Refitting the pedal shaft is the reverse of the removal procedure, following which it will be necessary to adjust the cable, as described in Section 18. On rhd variants check that the pedal has 0.24 to 0.55 in (6 to 14 mm) lift from the idle position; if necessary adjust the pedal lift-up stop to achieve this.

20 Exhaust system - general description

The exhaust system is of a two piece design consisting of a front downpipe and a length incorporating a front and rear muffler assembly. Running down the left-hand side of the car the system is supported by brackets at the front and rear mufflers.
When a part, or the complete exhaust system, other than the front pipe, is replaced for the first time a special sleeve and two 'U' clamps are necessary to connect the replacement sections of the front and rear muffler assembly. This is because the system is fitted whole in production and must be cut (to clear the rear suspension) to enable its removal.
The sleeve is obtainable from Ford dealers.

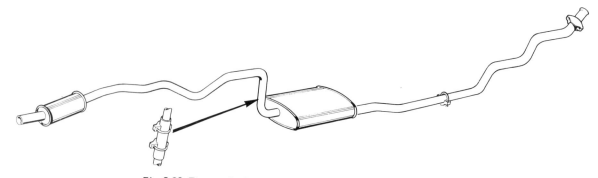

Fig. 3.29. The standard equipment system layout. The service sleeve is shown and the arrow indicates the fitting position (Sec. 20)

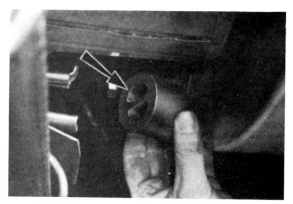

Fig. 3.30. Remove the mounting rubber (Sec. 21)

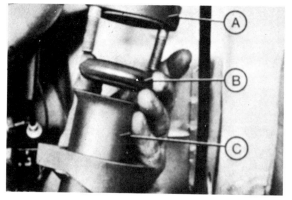

Fig. 3.31. Front pipe to manifold location (Sec. 21)

A Manifold   B Sealing ring   C Front pipe

Fig. 3.32. Mark the pipe at approximately 9½ in (241 mm) from the resonator box ...

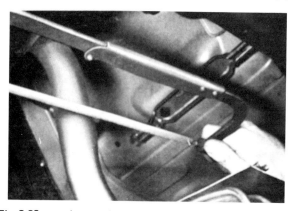

Fig. 3.33. ... and cut at the measured mark with a hacksaw (Sec. 21)

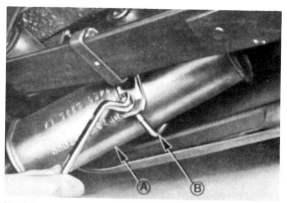

Fig. 3.34. Unscrew the muffler bracket retaining nut (Sec. 21)

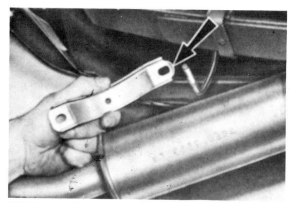

Fig. 3.35. The muffler support clamp. Arrow points to the angled end, which must be fitted to the top (Sec. 21)

Chapter 3/Carburation; fuel and exhaust systems

A regular check should be made on the exhaust system, for signs of leakage, secureness and general condition. Never allow a leaky exhaust to go unattended.

21 Exhaust pipe unit - removal and refitting

1 Ideally the car should be driven over a ramp or pit. If these are not available, jack up the left-hand side of the car and support with blocks. Chock the front wheel on the right-hand side and ensure that the handbrake is applied.
2 Raise up the resonator box and withdraw the mounting rubbers, Fig. 3.30.
3 Disconnect the manifold to front pipe retaining bolts and remove the sealing ring (Fig. 3.31). Use penetrating oil if necessary to free the bolts.
4 With a hacksaw, cut the exhaust pipe at the rear of the resonator as shown in Figs. 3.32 and 3.33. Then remove the front exhaust pipe section.
5 Unscrew the rear muffler bracket securing clamp nut (Fig. 3.34). and swing the bracket clear. Now remove the rear section of the pipe.
6 To replace the pipe, first measure and make a mark on the pipe at 45 mm (1.8 in) to the rear of the resonator.
7 Measure and mark at the same distance from the muffler inlet pipe end.
8 Fit the resonator and front pipe together and loosely fit the 'U' clamp.
9 With the resonator and muffler mounting brackets in position on the underbody, refit the mounting rubbers and refit the front pipe and resonator unit. Do not tighten the manifold and bracket connections at this stage.
10 Slide the service sleeve over the resonator pipe up to the scribed mark.
11 Now insert the muffler pipe into the service sleeve up to its mark and refit the rear mounting bracket to support but do not tighten yet. Note that the angled end of the clamp is fitted at the top (Fig. 3.35).
12 Check the exhaust system alignment and ensure that a minimum clearance of 25 mm (1.0 in) exists between the exhaust system and body or components.
13 Tighten the respective clamp and bracket bolts/nuts, and fit the 'U' clamps to the service sleeve (Fig. 3.36).
14 Note that the resonator bracket is tightened to the point where approximately ½ in (13 mm) of thread protrudes.
15 Start the engine and check for leaks.

Fig. 3.36. The exhaust system service sleeve in position; Note position of 'U' clamps (Sec. 21)

22 Muffler unit - removal and refitting

1 To remove and refit the muffler unit, follow the instructions given in Section 21, paragraphs 1, 4 to 7 and 10 to 13 respectively.
2 Check the system for leaks on reassembly.

23 Resonator unit - removal and refitting

1 To remove and replace the resonator and pipe unit, follow the instructions given in Section 21, paragraphs 1 to 4 and 6 to 8.
2 Slide the service sleeve over the muffler pipe up to the measured mark.
3 Fit the resonator and front pipe together and loosely fit the 'U' clamp.
4 With the resonator mounting bracket in position to the underbody, refit the mounting rubbers and reassemble the front pipe and resonator unit, inserting the exhaust end of the resonator pipe into the service sleeve, up to the measured mark.
5 Now follow the instructions given in Section 21, paragraphs 12 to 15 respectively.

Fault diagnosis overleaf

24 Fault diagnosis - carburation, fuel and exhaust systems

Symptom	Reason/s	Remedy
Fuel consumption excessive		
Carburation and ignition faults	Air cleaner choked and dirty giving rich mixture	Remove, clean and replace air cleaner.
	Fuel leaking from carburettor, fuel pump or fuel lines	Check for and eliminate all fuel leaks. Tighten fuel line union nuts.
	Float chamber flooding	Check and adjust float level.
	Generally worn carburettor	Remove, overhaul and replace.
	Distributor condenser faulty	Remove, and fit new unit.
	Balance weights or vacuum advance mechanism in distributor faulty	Remove, and overhaul distributor.
Incorrect adjustment	Carburettor incorrectly adjusted, mixture too rich	Tune and adjust carburettor.
	Idling speed too high	Adjust idling speed.
	Contact breaker gap incorrect	Check and reset gap.
	Valve clearances incorrect	Check rocker arm to valve stem clearances and adjust as necessary.
	Incorrectly set spark plugs	Remove, clean and regap.
	Tyres under-inflated	Check tyre pressures and inflate if necessary.
	Wrong spark plugs fitted	Remove and replace with correct units.
	Brakes dragging	Check and adjust brakes.
Insufficient fuel delivery or weak mixture due to air leaks		
Dirt in system	Petrol tank air vent restricted	Remove petrol tank and clean out air vent pipe.
	Partially clogged filters in pump and carburettor	Remove and clean filters.
	Dirt lodged in float chamber needle housing	Remove and clean out float chamber and needle valve assembly.
	Incorrectly seating valves in fuel pump	Remove, dismantle, and clean out fuel pump.
Fuel pump faults	Fuel pump diaphragm leaking or damaged	Remove, and overhaul fuel pump.
	Gasket in fuel pump damaged	Remove, and overhaul fuel pump.
	Fuel pump valves sticking due to petrol gumming	Remove, and thoroughly clean fuel pump.
Air leaks	Too little fuel in fuel tank (prevalent when climbing steep hills)	Refill fuel tank.
	Union joints on pipe connections loose	Tighten joints and check for air leaks.
	Split in fuel pipe on suction side of fuel pump	Examine, locate and repair.
	Inlet manifold to block or inlet manifold to carburettor gasket leaking	Test by pouring oil along joints - bubbles indicate leak. Renew gasket as appropriate.

Chapter 4 Ignition system

For modifications, and information applicable to later models, see Supplement at end of manual

Contents

Condenser - removal, testing and refitting ... 4	Fault diagnosis - ignition system ... 9
Contact breaker points - adjustment ... 2	General description ... 1
Contact breaker points - removal and refitting ... 3	Ignition timing ... 7
Distributor - reassembly and refitting ... 6	Spark plugs and HT leads ... 8
Distributor - removal, dismantling and inspection ... 5	

Specifications

Coil
Make ...	Motorcraft
Type ...	Low voltage with 1.5 ohm ballast resistor wire

Distributor
Make ...	Motorcraft
Type ...	Single pair contact breaker point
Automatic advance ...	Mechanical and vacuum
Drive ...	Skew gear from camshaft
Rotates ...	Anticlockwise (viewed from top)
Contact breaker points gap ...	Motorcraft 0.025 in (0.64 mm)
Dwell angle ...	48° to 52°
Distributor shaft endfloat ...	0.025 to 0.033 in (0.64 to 0.84 mm)
Static advance ...	6° BTDC

Spark plugs
Type ...	Motorcraft AGR22
Electrode gap ...	0.025 in (0.64 mm)

Torque wrench settings
	lb f ft	kg f m
Spark plugs ...	22 - 28	2.9 - 3.9

1 General description

In order that the engine can run correctly it is necessary for an electrical spark to ignite the fuel/air mixture in the combustion chamber at exactly the right moment in relation to engine speed and load. The ignition system is based on feeding low tension voltage from the battery to the coil where it is converted to high tension voltage. The high tension voltage is powerful enough to jump the spark plug gap in the cylinders many times a second under high compression, providing that the system is in good condition and that all adjustments are correct.

The ignition system is divided into two circuits, the low tension circuit and the high tension circuit.

The low tension (sometimes known as the primary) circuit consists of the battery, lead to the control box, lead to the ignition switch, lead from the ignition switch to the low tension or primary coil windings (terminal SW), and the lead from the low tension coil windings (coil terminal CB) to the contact breaker points and condenser in the distributor.

The high tension circuit consists of the high tension or secondary coil windings, the heavy ignition lead from the centre of the coil to the centre of the distributor cap, the rotor arm, and the spark plug leads and spark plugs.

The system functions in the following manner. Low tension voltage is changed in the coil into high tension voltage by the opening and closing of the contact breaker points in the low tension circuit. High tension voltage is then fed via the carbon brush in the centre of the distributor cap to the rotor arm of the distributor cap, and each time it comes in line with one of the four metal segments in the cap, which are connected to the spark plug leads, the opening and closing of the contact breaker points causes the high tension voltage to build up, jump the gap from the rotor arm to the appropriate metal segment and so via the spark plug lead to the spark plug, where it finally jumps the spark plug gap before going to earth.

The ignition is advanced and retarded automatically, to ensure that the spark occurs at just the right instant for the particular load at the prevailing engine speed.

The ignition advance is controlled both mechanically and by a vacuum operated system. The mechanical governor mechanism comprises two weights, which move out from the distributor shaft as the engine speed rises due to centrifugal force. As they move outwards they rotate the cam relative to the distributor shaft, and so advance the spark. The weights are held in position by two light springs and it is the tension of the springs which is largely responsible for correct spark advancement.

The vacuum control consists of a diaphragm, one side of which is connected via a small bore tube to the carburettor, and the other side to the contact breaker plate. Depression in the inlet manifold and carburettor, which varies with engine speed and throttle opening, causes the diaphragm to move, so moving the contact breaker plate, and advancing or retarding the spark. A spring located in the vacuum

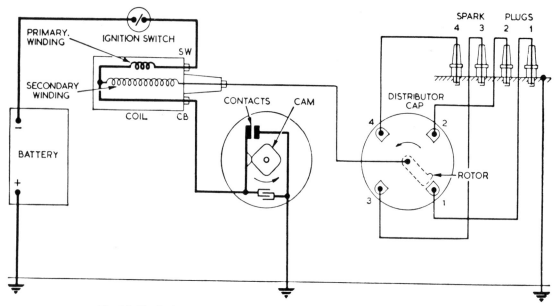

Fig. 4.1. The ignition circuit - heavier line indicates LT (primary) circuit (Sec. 1)

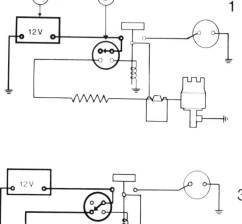

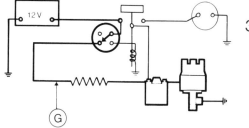

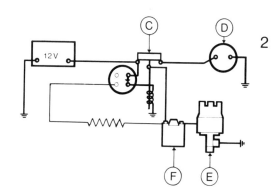

Fig. 4.2. The three stages of the ignition circuit (Sec. 1)

1. The ignition switched off
2. The ignition switched in the start position
3. The ignition switched in the on position

A Battery
B Ignition switch
C Starter solenoid
D Starter motor
E Distributor
F Ignition coil
G Ballast resistor wire

unit ensures that the breaker plate returns to the normal position when the manifold depression decreases.

The wiring harness incorporates a resistance wire in the ignition coil feed circuit and it is vital that an ignition coil with a ballast resistor is used. The starter solenoid has an extra terminal in order that the wire from the solenoid to the coil supplies voltage direct to the coil when the starter motor is operated. The ballast resistor is therefore bypassed and battery voltage is fed to the ignition system to give easier starting.

On cars fitted with an FM radio, a screening can is fitted round the distributor to suppress interference. This can easily be removed for access to the distributor.

2 Contact breaker points - adjustment

1   To adjust the contact breaker points to the correct gap, first pull off the two clips securing the distributor cap to the distributor body, and lift away the cap. Clean the cap inside and out with a dry cloth. It is unlikely that the four segments will be badly burned or scored, but if they are the cap will have to be renewed.

2   Inspect the carbon brush contact located in the top of the cap - see that it is unbroken and stands proud of the plastic surface.

3   Check the contact spring on the top of the rotor arm. It must be clean and have adequate tension to ensure good contact.

4   Gently prise the contact breaker points open to examine the condition of their faces. If they are rough, pitted or dirty, it will be necessary to remove them for resurfacing, or for new points to be fitted.

5   Assuming the points are satisfactory, or that they have been cleaned and replaced, measure the gap between the points by turning the engine over until the heel of the breaker arm is on the highest point of the cam.

6   An 0.025 inch (0.64 mm) feeler gauge should now just fit between the points.

7   If the gap varies from this amount slacken the contact plate securing screw.

8   Adjust the contact gap by inserting a screwdriver in the notched hole in the breaker plate. Turn clockwise to increase and anticlockwise to decrease the gap. When the gap is correct, tighten the securing screw

Chapter 4/Ignition system

Fig. 4.3. Contact breaker gap measuring point (A) (Sec. 2)

Note that the heel of the point assembly rests on the high point of cam

Fig. 4.4. Contact breaker removal (Sec 3)

Screws at (A) retain assembly to base plate, screw (B) holds condenser and LT lead to assembly

Fig. 4.5. Contact breaker point assembly retaining screws (Sec. 3)

Fig. 4.6. Greasing the cam lobes (A) (Sec. 3)

and check the gap again.

9 Make sure the rotor is in position. Replace the distributor cap and clip the spring blade retainers into position.

3 Contact breaker points - removal and refitting

1 If, on inspection, the faces of the contacts are burnt, pitted or worn, the contact breaker points must be removed for refacing or renewal.
2 Lift off the rotor arm by pulling it straight up from the spindle.
3 Slacken the self-tapping screw holding the condenser and low tension leads to the contact breaker and slide out the forked ends of the leads.
4 Remove the points by taking out the two retaining screws and lifting off the points assembly.
5 Dress the face of each contact squarely on an oilstone or a piece of fine emery cloth until all traces of 'pips' or 'craters' have been removed. After two or three times, regrinding of the points in this manner will reduce the thickness of the metal so much that a new contact set will have to be fitted. Before fitting the points, clean them with methylated spirit.
6 Refitting the points assembly is a reversal of removal but take care not to trap the wires between the points and the contact breaker plate.
7 Set the points gap as described in the preceding Section.
8 Refit the rotor arm and the distributor cap.
9 Whenever the contact breaker points are serviced or adjusted, the distributor should be lubricated. Smear the high points of the cam (Fig. 4.6) with petroleum jelly and apply two or three drops of engine oil to the felt pad which is located at the top of the cam assembly (Fig. 4.6). Squirt a few drops of engine oil through the distributor baseplate to lubricate the mechanical advance and retard assembly. Do not lubricate the distributor too liberally otherwise the points will become contaminated and misfiring will occur.

4 Condenser - removal, testing and refitting

1 The purpose of the condenser (capacitor) is to ensure that when the contact breaker points open, there is no sparking across them which would waste voltage and cause wear.
2 The condenser is fitted in parallel with the contact breaker points. If it develops a short circuit, it will cause ignition failure as the points will be prevented from interrupting the low tension circuit.
3 If the engine becomes very difficult to start or begins to miss after several miles running and the breaker points show signs of excessive burning, then the condition of the condenser must be suspect. A further test can be made by separating the points by hand with the ignition switched on. If this is accompanied by a flash it is indicative that the condenser has failed.
4 Without special test equipment the only sure way to diagnose condenser trouble is to replace a suspected unit with a new one and note if there is any improvement.
5 To remove the condenser from the distributor, take off the distributor cap and rotor arm. Slacken the self-tapping screw holding the condenser lead and low tension lead to the points, and slide out the fork on the condenser lead. Undo the condenser retaining screw and remove the condenser from the breaker plate (Fig. 4.7).
6 To refit the condenser simply reverse the order of removal. Take care that the condenser lead is clear of the moving part of the points assembly.

5 Distributor - removal, dismantling and inspection

1 To remove the distributor from the engine, pull off the four leads from the spark plugs, tagging each one to ensure correct replacement.
2 Disconnect the high and low tension leads from the distributor and then remove the distributor cap.

Chapter 4/Ignition system

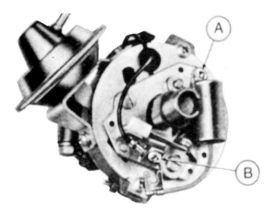

Fig. 4.7. Condenser removal

Screw (A) retains condenser to base plate, screw (B) retains condenser lead to contact breaker assembly (Sec. 4)

Fig. 4.8. Engine and rotor positions for TDC on No. 1 cylinder (Sec. 5)

A Rotor arm adjacent to No. 1 contact
B Timing notch on pulley next to TDC mark on front cover

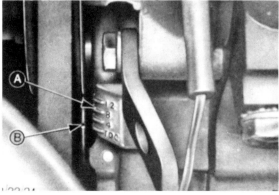

Fig. 4.9. Crankshaft pulley timing notch (B) and 8° BTDC mark (A) shown arrowed. Note that as shown engine is in TDC position on No. 1 or 4 cylinder (Sec. 5)

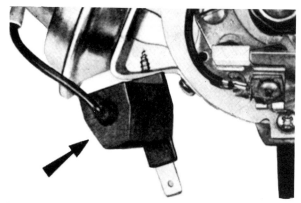

Fig. 4.10. Radio suppressor location on vacuum bracket (Sec. 5)

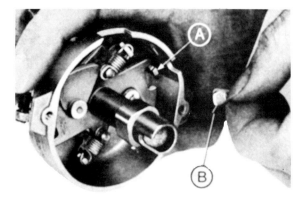

Fig. 4.11. Removing the bump stop plastic cover (B) from bump stop (A) (Sec. 5)

Fig. 4.12. Base plate removed - the vacuum unit pivot post (A) circlip must be removed together with base plate retaining screws (D). Also arrowed is screw (C) holding the radio suppressor to the vacuum unit bracket (Sec. 5)

3 Pull off the rubber union holding the vacuum pipe to the distributor vacuum advance housing.
4 Using a suitably sized spanner on the crankshaft pulley bolt turn the engine to TDC on No. 1 cylinder. This can be readily achieved by noting the position of the rotor arm; when the rotor arm reaches the point where it would normally be opposite the No. 1 contact in the distributor cap No. 1 cylinder is approaching TDC. Continue turning the engine until the notch in the crankshaft pulley aligns with the appropriate timing mark on the front cover (Fig. 4.8).
5 Remove the distributor body clamp bolt which holds the distributor clamp plate to the cylinder block. Pull the distributor from the block, marking the position of the rotor arm relative to the body.
6 Pull the rotor arm off the distributor camshaft. Remove the points from the breaker plate as detailed in Section 3.
7 Undo the condenser retaining screw and take off the condenser.
8 Next prise off the small circlip from the vacuum unit pivot post.

Refer to Fig. 4.12.
9 Take out the two screws holding the breaker plate to the distributor body and lift away. Where a vehicle is fitted with a VHF (FM) radio there may be a suppressor fitted to the vacuum unit bracket, Fig. 4.10. Remove the suppressor by unscrewing the single screw holding it to the bracket.
10 Take off the circlip, flat washer and wave washers from the pivot post. Separate the two plates by bringing the holding down screw through the keyhole slot in the lower plate. Be careful not to lose the spring now left on the pivot post.
11 Pull the low tension wire and grommet from the lower plate.
12 Undo the two screws holding the vacuum unit to the body. Take off the unit (Fig. 4.15).
13 Prise the plastic bump stop tubing off the advance stop in the distributor body (Fig. 4.11).
14 The mechanical advance is next removed but first make a careful

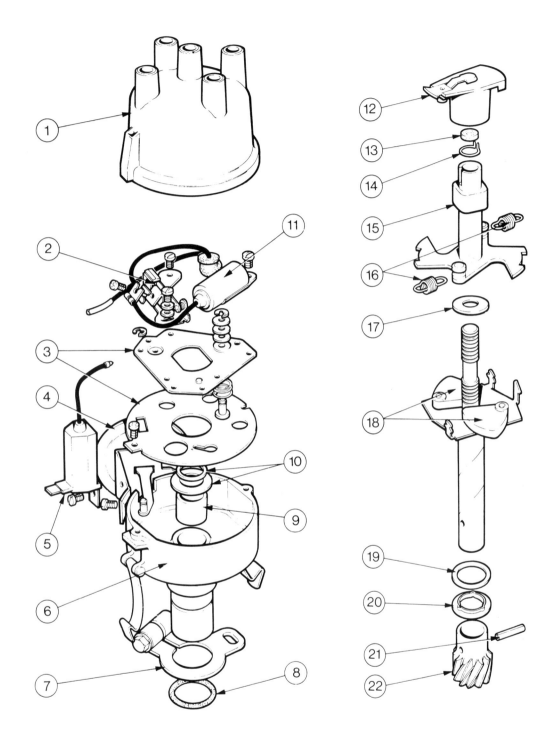

Fig. 4.13. The component parts of the distributor

1 Cap
2 Contact breaker assembly
3 Base plate
4 Vacuum unit
5 Suppressor
6 Distributor body
7 Body clamp plate
8 Cylinder block seal
9 Bush (not serviced)
10 Thrust washers
11 Condenser
12 Rotor arm
13 Felt pad
14 Circlip
15 Cam spindle
16 Mechanical advance springs
17 Washer
18 Mechanical advance weights
19 Washer
20 Washer
21 Pin
22 Gear

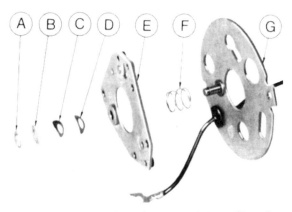

Fig. 4.14. Distributor base plate - exploded view (Sec. 5)

Fig. 4.15. Screws (arrowed) retaining vacuum unit to distributor (Sec. 5)

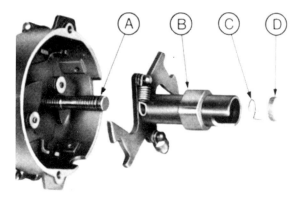

Fig. 4.16. Cam spindle assembly (Sec. 5)

A Shaft
B Cam spindle
C Circlip
D Felt pad

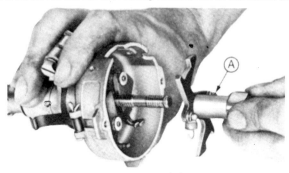

Fig. 4.17. Withdrawing cam spindle (A) from distributor body

6 Distributor - reassembly and refitting

1 Reassembly is a reversal of dismantling but observe the following points:
2 Apply a film of engine oil to the shaft and mechanical advance assembly before fitting.
3 When fitting the lower shaft, first replace the thrust washers below the action plate before inserting into the distributor body. Next fit the wave washer and thrust washer at the lower end and replace the drive gear. Secure it with a new pin.
4 Assemble the upper and lower shaft with the advance stop in the correct slot (the one which was marked) in the mechanical advance plate.
5 After assembling the advance weights and springs check that they move freely without binding, then replace the advance stop tubing.
6 Before assembling the breaker plates, make sure that the three nylon bearing studs are properly located in their holes in the upper breaker plate, and that the small earth spring is fitted on the pivot post.
7 As you refit the upper breaker plate, pass the holding down spindle through the keyhole slot in the lower plate.
8 Hold the upper plate in position by refitting the wave washer, flat washer and large circlip.
9 When all is assembled, remember to set the contact breaker gap to 0.025 inch (0.64 mm).
10 If a new gear or shaft is being fitted, it is necessary to drill a new pin hole. Proceed as follows.
11 Make a 0.015 inch (0.38 mm) thick forked shim to slide over the driveshaft (Fig. 4.19).
12 Assemble the shaft, wave washer, thrust washer, shim and gear wheel in position in the distributor body.
13 Hold the assembly in a large clamp such as a vice or carpenter's clamp using only sufficient pressure to take up all end play.
14 There is a pilot hole in the new gear wheel for drilling the new hole. Set this pilot hole at 90° to the existing hole in an old shaft if the old shaft is being re-used. Drill an 1/8 inch (3.18 mm) hole through both gear and shaft.
15 Fit a new pin in the hole. Release the clamp and remove the shim. The shaft will now have the correct amount of clearance.

note of the assembly, particularly which spring fits which post and the position of the advance springs. Then remove the advance springs.
15 Prise off the circlips from the governor weight pivot pins and take out the weights.
16 Dismantle the shaft by taking out the felt pad in the top of the spindle. Expand the exposed circlip and take it out.
17 Now mark which slot in the mechanical advance plate is occupied by the advance stop which stands up from the action plate, and lift off the cam spindle (Fig. 4.17).
18 It is only necessary to remove the lower shaft and action plate if it is excessively worn. If this is the case, with a small punch drive out the gear retaining pin and remove the gear with the two washers located above it. Withdraw the shaft from the distributor body and take off the two washers from below the action plate. The distributor is now completely dismantled.
19 Check the points, as described in Section 3. Check the distributor cap for signs of tracking, indicated by a thin black line between the segments. Renew the cap if any signs of tracking are found.
20 If the metal portion of the rotor arm is badly burned or loose, renew the arm. If only slightly burned, clean the end with a fine file. Check that the contact spring has adequate pressure and the bearing surface is clean and in good condition.
21 Check that the carbon brush in the distributor cap is unbroken and stands proud of its holder.
22 Examine the fly weights and pivots for wear and the advance springs for slackness. They can best be checked by comparing with new parts. If they are slack they must be renewed.
23 Check the points assembly for fit on the breaker plate, and the cam spindle for wear.
24 Examine the fit of the lower shaft in the distributor body. If this is excessively worn, it will be necessary to fit a new assembly.

Chapter 4/Ignition system

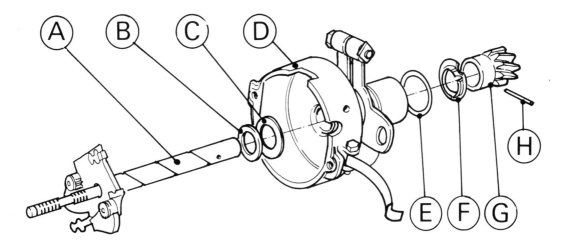

Fig. 4.18. Distributor shaft assembly (Sec. 6)

A Shaft
B Thrust washer
C Thrust washer
D Distributor body
E Wave washer
F Washer
G Gear
H Pin

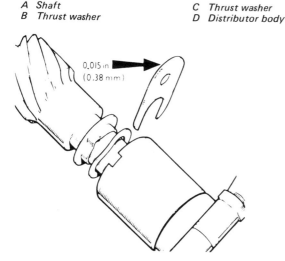

Fig. 4.19. Shaft endfloat adjustment plate (Sec. 6)

Fig. 4.20. Slacken the distributor clamp pinch bolt (Sec. 6)

16 Hold the distributor above its crankcase opening so that the vacuum advance unit is pointing to the rear of the car and set the contact end of the rotor arm to align with No. 2 inlet port. Insert the distributor; as its driveshaft meshes with the camshaft gear the rotor will rotate slightly and take up a position of approximately 90° to the centre-line of the engine.
17 Fit the clamp plate retaining bolt to hold the assembly to the engine block and tighten it.
18 Slacken the distributor clamp pinch bolt (Fig. 4.20).
19 Gently turn the distributor body until the contact breaker points are just opening when the rotor is pointing at the contact in the distributor cap which is connected to No. 1 spark plug. A convenient way is to put a mark on the outside of the distributor body in line with the terminal cover, so that it shows when the cover is removed.
20 If this position cannot easily be reached, check that the drive gear has meshed on the correct tooth by lifting out the distributor once more. If necessary rotate the driveshaft one tooth and try again.
21 Tighten the distributor body clamp enough to hold the distributor, but do not overtighten.

7 Ignition timing

1 The basic procedure for timing the ignition is fully described in the preceding Section.
2 The precise opening point of the contact breaker points can best be checked by connecting a test bulb between the distributor LT terminal and a good earth. Switch on the ignition and rotate the distributor until the lamp just lights; this indicates that the points are just opening.
3 If the ignition timing is to be checked by stroboscope, the manufacturer's instructions should be followed. Always remove the vacuum pipe from the distributor before testing by this method.
4 Since the ignition timing setting enables the firing point to be correctly related to the grade of fuel used, the fullest advantage of a change of grade from that recommended for the engine will only be attained by re-adjustment of the ignition setting.

8 Spark plugs and HT leads

1 The correct functioning of the spark plugs is vital for the correct running and efficiency of the engine.
2 At intervals of 6,000 miles (9,600 km) the plugs should be removed, examined, cleaned, and if worn excessively, renewed. The condition of the spark plugs will also tell much about the overall condition of the engine (see illustrations on page 65).
3 If the insulator nose of the spark plug is clean and white, with no deposits, this is indicative of a weak mixture, or too hot a plug. (A hot plug transfers heat away from the electrode slowly - a cold plug transfers it away quickly).
4 The plugs fitted as standard are Autolite as listed in the Specifications at the beginning of this Chapter. If the tip and insulator nose is covered with hard black looking deposits, then this is indicative that the mixture is too rich. Should the plug be black and oily, then it is likely that the engine is fairly worn, as well as the mixture being too rich.

5 If the insulator nose is covered with light tan to greyish brown deposits, then the mixture is correct and it is likely that the engine is in good condition.

6 If there are any traces of long brown tapering stains on the outside of the white portion of the plug, then the plug will have to be renewed, as this shows that there is a faulty joint between the plug body and the insulator, and compression is being lost.

7 Plugs should be cleaned by a sand blasting machine, which will free them from carbon more thoroughly than cleaning by hand. The machine will also test the condition of the plugs under compression. Any plug that fails to spark at the recommended pressure should be renewed.

8 The spark plug gap is of considerable importance, as, if it is too large or too small, the size of the spark and its efficiency will be seriously impaired. The spark plug gap should be set to the figure given in the Specifications at the beginning of this Chapter.

9 To set it, measure the gap with a feeler gauge, and then bend open, or close, the **outer** plug electrode until the correct gap is achieved. The centre electrode should **never** be bent as this may crack the insulation and cause plug failure if nothing worse.

10 When replacing the plugs, remember to replace the leads from the distributor in the correct firing order, which is 1, 2, 4, 3, No. 1 cylinder being the one nearest the radiator. No. 1 lead from the distributor runs from the one o'clock position when looking down on the distributor cap, 2, 4 and 3 are anticlockwise from No. 1. (Fig. 4.21).

11 The plug leads require no routine attention other than being kept clean and wiped over regularly. At intervals of 6,000 miles (9,600 km), however, pull the leads off the plugs and distributor one at a time and make sure no water has found its way onto the connections. Remove any corrosion from the brass ends, wipe the collars on top of the distributor and refit the leads.

12 Every 10,000 to 12,000 miles (16,000 to 19,000 km) it is recommended that the spark plugs are renewed to maintain optimum engine performance.

13 Later vehicles are fitted with carbon cored HT leads. These should be removed from the spark plugs by gripping their crimped terminals. Provided the leads are not bent in a tight loop carbon cored there is no reason why this type of lead should fail. A legend has arisen which blames this type of lead for all ignition faults and many owners replace them with the older copper cored type and install separate suppressors. In the majority of cases, it would be more profitable to establish the real cause of the trouble before going to the expense of new leads.

9 Fault diagnosis - ignition system

By far the majority of breakdowns and running troubles are caused by faults in the ignition system, either in the low tension or high tension circuit. There are two main symptoms indicating ignition faults. Either the engine will not start or fire, or the engine is difficult to start and misfires. If it is a regular misfire, ie. the engine is only running on two or three cylinders, the fault is almost sure to be in the secondary, or high tension, circuit. If the misfiring is intermittent, the fault could be in either the high or low tension circuits. If the car stops suddenly, or will not start at all, it is likely that the fault is in the low tension circuit. Loss of power and overheating, apart from faulty carburation settings, are normally due to faults in the distributor or incorrect ignition timing.

Engine fails to start

1 If the engine fails to start and the car was running normally when it was last used, first check there is fuel in the petrol tank. If the engine turns over normally on the starter motor and the battery is evidently well charged, then the fault may be in either the high or low tension circuits. First check the HT circuit. **Note:** if the battery is known to be fully charged; the ignition light comes on, and the starter motor fails to turn the engine, **check the tightness of the leads on the battery terminals** and the security of the earth lead to its **connection on the body.** It is quite common for the leads to have worked loose, even if they look and feel secure. If one of the battery terminal posts gets very hot when trying to work the starter motor, this is a sure indication of a faulty connection to that terminal.

2 One of the commonest reasons for bad starting is wet or damp spark plug leads and distributor. Remove the distributor cap. If condensation is visible internally dry the cap with a rag and wipe over the leads.

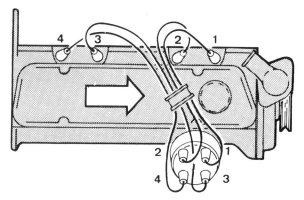

Fig. 4.21. The high tension plug leads to distributor layout in the firing order (Sec. 8)

Replace the cap.

3 If the engine still fails to start, check that current is reaching the plugs by disconnecting each plug lead in turn at the spark plug and holding the end of the cable about 3/16 inch (5 mm) away from the cylinder block. Spin the engine on the starter motor.

4 Sparking between the end of the cable and the block should be fairly strong with a regular blue spark. (Hold the lead with rubber to avoid electric shocks). If current is reaching the plugs, then remove them, and clean and regap them. The engine should now start.

5 If there is no spark at the plug leads, take off the HT lead from the centre of the distributor cap and hold it to the block as before. Spin the engine on the starter once more. A rapid succession of blue sparks between the end of the lead and the block indicates that the coil is in order and that the distributor cap is cracked, the rotor arm faulty or the carbon brush in the top of the distributor cap is not making good contact with the spring on the rotor arm. Possibly the points are in bad condition. Clean and reset them.

6 If there are no sparks from the end of the lead from the coil, check the connections at the coil end of the lead. If it is in order start checking the low tension circuit.

7 Use a 12 volt voltmeter or a 12 volt bulb and two lengths of wire. With the ignition switch on and the points open test between the low tension wire to the coil (it is marked SW or +) and earth. No reading indicates a break in the supply from the ignition switch. Check the connections at the switch to see if any are loose. Refit them and the engine should run. A reading shows a faulty coil or condenser or broken lead between the coil and the distributor.

8 Take the condenser wire off the points assembly and with the points open, test between the moving point and earth. If there now is a reading, then the fault is in the condenser. Fit a new one and the fault will be cleared.

9 With no reading from the moving point to earth, take a reading between earth and the CB or (−) terminal of the coil. A reading here indicates a broken wire which must be renewed between the coil and distributor. No reading confirms that the coil has failed and must be renewed. Remember to connect the condenser wire to the points assembly. For these tests it is sufficient to separate the contact breaker points with a piece of paper.

Engine misfires

1 If the engine misfires regularly, run it at a fast idling speed. Pull off each of the plug caps in turn and listen to the note of the engine. Hold the plug cap in a dry cloth or with a rubber glove as additional protection against a shock from the HT supply.

2 No difference in engine running will be noticed when the lead from the defective circuit is removed. Removing the lead from one of the good cylinders will accentuate the misfire.

3 Remove the plug lead from the end of the defective plug and hold it about 3/16 inch (5 mm) away from the block. Restart the engine. If the sparking is fairly strong and regular, the fault must lie in the spark plug.

4 The plug may be loose, the insulation may be cracked, or the points may have burnt away, giving too wide a gap for the spark to jump. Worse still, one of the points may have broken off. Either renew the plug, or clean it, reset the gap, and then test it.

5 If there is no spark at the end of the plug lead, or if it is weak and

Measuring plug gap. A feeler gauge of the correct size (see ignition system specifications) should have a slight 'drag' when slid between the electrodes. Adjust gap if necessary

Adjusting plug gap. The plug gap is adjusted by bending the earth electrode inwards, or outwards, as necessary until the correct clearance is obtained. Note the use of the correct tool

Normal. Grey-brown deposits, lightly coated core nose. Gap increasing by around 0.001 in (0.025 mm) per 1000 miles (1600 km). Plugs ideally suited to engine, and engine in good condition

Carbon fouling. Dry, black, sooty deposits. Will cause weak spark and eventually misfire. Fault: over-rich fuel mixture. Check: carburettor mixture settings, float level and jet sizes; choke operation and cleanliness of air filter. Plugs can be re-used after cleaning

Oil fouling. Wet, oily deposits. Will cause weak spark and eventually misfire. Fault: worn bores/piston rings or valve guides; sometimes occurs (temporarily) during running-in period. Plugs can be re-used after thorough cleaning

Overheating. Electrodes have glazed appearance, core nose very white – few deposits. Fault: plug overheating. Check: plug value, ignition timing, fuel octane rating (too low) and fuel mixture (too weak). Discard plugs and cure fault immediately

Electrode damage. Electrodes burned away; core nose has burned, glazed appearance. Fault: pre-ignition. Check: as for 'Overheating' but may be more severe. Discard plugs and remedy fault before piston or valve damage occurs

Split core nose (may appear initially as a crack). Damage is self-evident, but cracks will only show after cleaning. Fault: pre-ignition or wrong gap-setting technique. Check: ignition timing, cooling system, fuel octane rating (too low) and fuel mixture (too weak). Discard plugs, rectify fault immediately

intermittent, check the ignition lead from the distributor to the plug. If the insulation is cracked or perished, renew the lead. Check the connections at the distributor cap.

6 If there is still no spark, examine the distributor cap carefully for tracking. This can be recognised by a very thin black line running between two or more electrodes, or between an electrode and some other part of the distributor. These lines are paths which now conduct electricity across the cap, thus letting it run to earth. The only answer is a new distributor cap.

7 Apart from the ignition timing being incorrect, other causes of misfiring have already been dealt with under the Section dealing with the failure of the engine to start. To recap, these are that:

a) The coil may be faulty giving an intermittent misfire.
b) There may be a damaged wire or loose connection in the low tension circuit.
c) The condenser may be short circuiting.
d) There may be a mechanical fault in the distributor (broken driving spindle or contact breaker spring).

8 If the ignition timing is too far retarded, it should be noted that the engine will tend to overheat, and there will be a quite noticeable drop in power. If the engine is overheating and the power is down and the ignition timing is correct, then the carburettor should be checked, as it is likely that this is where the fault lies.

Chapter 5 Clutch

For modifications, and information applicable to later models, see Supplement at end of manual

Contents

Clutch adjustment ... 2	Clutch pedal - removal and refitting ... 8
Clutch - inspection and overhaul ... 4	Clutch release bearing - removal and refitting ... 6
Clutch - refitting ... 5	Fault diagnosis - clutch ... 9
Clutch - removal ... 3	General description ... 1
Clutch cable - removal and refitting ... 7	

Specifications

Manufacturer ...	Borg and Beck or Laycock
Type ...	Single dry plate, diaphragm spring
Actuation ...	Cable
Diameter - driven plate (friction disc) ...	7.5 in (190.5 mm)
Clutch lining ...	Mintex H26
Lining thickness ...	0.13 in (3.25 mm)
Effective clutch lining area ...	40.7 in (26221 mm)
Clutch plate spring pressure ...	847 lb (385 kg)
Clutch pedal free-travel ...	0.87 ± 0.16 in (22.0 ± 4.0 mm)
Release bearing type ...	Sealed ball

Torque wrench settings

	lbf ft	kgf m
Clutch pressure plate to flywheel bolts ...	12 to 15	1.66 to 2.07
Clutch housing to engine ...	22 to 27	3.0 to 3.7
Clutch housing to transmission ...	31 to 35	4.2 to 4.8

1 General description

The model covered by this manual is fitted with a 7.5 in (190.5 mm) clutch of either Borg and Beck or Laycock manufacture.

The clutch assembly comprises a steel cover which is dowelled and bolted to the rear face of the flywheel and contains the pressure plate, diaphragm spring and fulcrum rings.

The clutch disc is free to slide along the splined gearbox first motion shaft and is held in position between the flywheel and the pressure plate by the pressure of the pressure plate spring. Friction lining material is riveted to the clutch disc and it has a spring cushioned hub to absorb transmission shocks and to help ensure a smooth take-off.

The circular diaphragm spring is mounted on shouldered pins and held in place in the cover by two fulcrum rings. The spring is also held to the pressure plate by three spring steel clips which are riveted in position.

The clutch is actuated by a cable controlled by the clutch pedal. The clutch release mechanism consists of a release fork and bearing which are in permanent contact with the release fingers on the pressure plate. There should, therefore, never be any free play at the release fork. Wear of the friction material on the clutch disc is taken up by means of a cable adjuster on the clutch bellhousing.

Depressing the clutch pedal actuates the clutch release arm by means of the cable.

The release arm pushes the release bearing forward to bear against the release fingers, so moving the centre of the diaphragm spring inward. The spring is sandwiched between two annular rings which act as fulcrum points. As the centre of the spring is pushed in, the outside of the spring is pushed out, so moving the pressure plate backward and disengaging the pressure plate from the clutch disc.

When the clutch pedal is released, the diaphragm spring forces the pressure plate into contact with the friction linings on the clutch disc and at the same time pushes the clutch disc a fraction of an inch forward on its splines and engages with the flywheel. The clutch disc is now firmly sandwiched between the pressure plate and the flywheel and the drive taken-up.

2 Clutch - adjustment

1 Every 6000 miles (9600 km) the clutch pedal backlift (Fig. 5.2) must be checked and adjusted, if necessary, to compensate for wear in the friction linings. First, jack up the front of the car and support on axle stands or blocks. Then chock the rear wheels.

2 From underneath the car, slacken the locknut on the threaded portion of the outer cable at the clutch bellhousing.

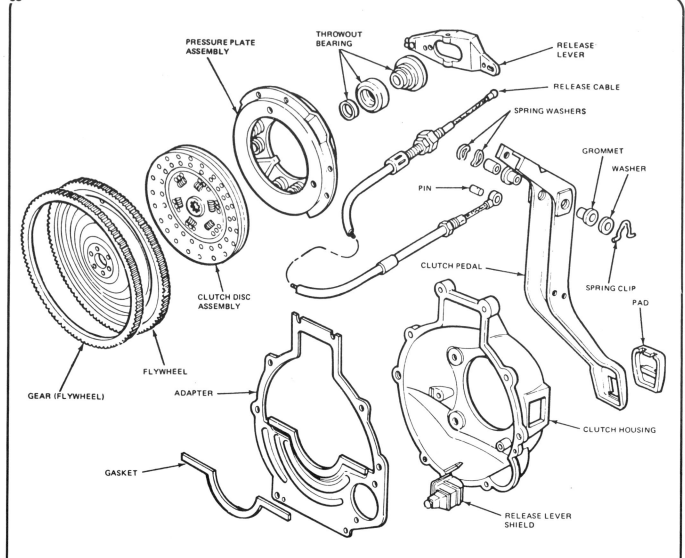

Fig. 5.1. The clutch assembly component parts

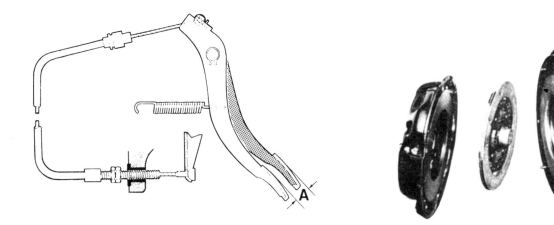

Fig. 5.2. Clutch pedal free travel adjustment: BACKLIFT at 'A' on the pedal pad must be 22.0 ± 4.0 mm (0.87 ± 0.16 in) (Sec. 2)

Fig. 5.3. The clutch removed from the flywheel (Sec. 3)

3 Check that the cable is not kinked or frayed, and then grasp the outer case of the cable and pull the cable forward to take up any free play in the cable.
4 Turn the adjusting nut down the adjuster thread until it contacts the cable bush in the bellhousing, then press the pedal to the floor several times to ensure all components are properly seated.
5 Reset the adjusting nut until there is a free backlift movement of 22.0 ± 4.0 mm at the clutch pedal pad. Before checking the backlift push the pedal slowly down to the floor and, equally slowly, allow it to return to the 'rest' position. Allowing the pedal to spring back to the 'rest' position may not necessarily cause it to reach its normal 'rest' position.
6 If the pedal backlift movement is too small *slacken* the adjusting nut, if too great *tighten* the adjusting nut.
When the adjustment is correct tighten the locknut without disturbing the adjusting nut, then recheck the backlift.
7 Lower the car and remove the jack and rear wheel chocks.

3 Clutch - removal

1 Remove the gearbox as described in Chapter 6, or alternatively the engine as described in Chapter 1. Although the removal of the gearbox is the easier of the two methods, a considerable amount of working space is required beneath the car and ideally, ramps or an inspection pit should be used. However, provided that suitable jacks and supports are available, and the car can be raised high enough in complete safety, then the gearbox can be removed without a pit or ramps.
2 Scribe a mating line from the clutch cover to the flywheel to ensure identical positioning on replacement and then remove the clutch assembly by unscrewing the six bolts holding the cover to the rear face of the flywheel. Unscrew the bolts diagonally half a turn at a time to prevent distortion of the cover flange.
3 With all the bolts and spring washers removed, lift the clutch assembly off the locating dowels. The clutch disc may fall out at this stage as it is not attached to either the clutch cover assembly or the flywheel.

4 Clutch - inspection and overhaul

1 It is not practical to dismantle the pressure plate assembly. If a new clutch disc is being fitted it is false economy not to renew the release bearing at the same time. This will preclude having to replace it at a later date when wear on the clutch linings is very small.
2 If the pressure plate assembly requires renewal an exchange unit must be purchased. This will have been accurately set up and balanced to very fine limits.
3 Examine the clutch disc friction linings for wear and loose rivets and the plate for rim distortion, cracks, broken hub springs, and worn splines. The surface of the friction linings may be highly glazed, but as long as the clutch material pattern can be clearly seen this is satisfactory. Compare the amount of lining wear with a new clutch disc at the stores in your local garage, and if the linings are more than three quarters worn renew the disc.
4 It is always best to renew the clutch disc as an assembly to preclude further trouble, but, if it is wished to merely renew the linings, the rivets should be drilled out and not knocked out with a punch. The manufacturers do not advise that only the linings are renewed and personal experience dictates that it is far more satisfactory to renew the clutch disc complete than to try and economise by fitting only new friction linings.
5 Check the machined faces of the flywheel and the pressure plate. If either is grooved it should be machined until smooth or renewed.
6 If the pressure plate is cracked or split it is essential that an exchange unit is fitted, also if the pressure of the diaphragm spring is suspect.
7 Check the release bearing for smoothness of operation. There should be no harshness and no slackness in it. It should spin freely, bearing in mind it has been pre-packed with grease. If renewal is necessary, refer to Section 6.
8 Check also that the clutch pilot bearing in the centre of the flywheel is serviceable. Further information on this will be found in Chapter 1, Section 28.

Fig. 5.4. Centralising the clutch driven plate (Sec. 5)

5 Clutch - refitting

1 It is important that no oil or grease gets on the clutch disc friction linings, or the pressure plate and flywheel faces. It is advisable to replace the clutch with clean hands and to wipe down the pressure plate and flywheel faces with a clean dry rag before reassembly begins.
2 Place the clutch disc against the flywheel, ensuring that it is the correct way round. The flywheel side of the clutch disc is smooth and the hub boss is longer on this side. If the disc is fitted the wrong way round, it will be quite impossible to operate the clutch.
3 Refit the clutch cover assembly loosely on the dowel and ensure that the scribed marks on the flywheel cover align. Refit the six bolts and spring washers, and tighen them finger tight so that the clutch disc is gripped but can still be moved.
4 The clutch disc must now be centralised so that when the engine and gearbox are mated, the gearbox input shaft splines will pass through the splines in the centre of the driven plate hub.
5 Centralisation can be carried out quite easily by inserting a round bar or long screwdriver through the hole in the centre of the clutch, so that the end of the bar rests in the small hole in the end of the crankshaft containing the input shaft pilot bush. Ideally an old input shaft should be used.
6 Using the input shaft pilot bush as a fulcrum and moving the bar sideways or up and down, will move the clutch disc in whichever direction is necessary to achieve centralisation.
7 Centralisation is easily judged by removing the bar and moving the disc hub in relation to the hole in the centre of the clutch cover diaphragm spring. When the hub appears exactly in the centre of the hole all is correct. Alternatively the input shaft will fit the bush and centre of the clutch hub exactly obviating the need for visual alignment.
8 Tighten the clutch bolts firmly in a diagonal sequence to ensure that the cover plate is pulled down evenly and without distortion of the flange. Finally, tighten the bolts down to the specified torque.

6 Clutch release bearing - removal and refitting

1 With the gearbox and engine separated to provide access to the clutch, attention can be given to the release bearing located in the bellhousing, over the input shaft.
2 The release bearing is a relatively inexpensive but important component and unless it is nearly new it is a mistake not to renew it during an overhaul of the clutch.
3 To remove the release bearing first pull off the release arm rubber gaiter.
4 The release arm and bearing assembly can then be withdrawn from the clutch housing.
5 To free the bearing from the release arm simply unhook it, and then with the aid of two blocks of wood and a vice press off the release bearing from its hub.
6 Refitting is a reversal of removal, but lubricate the hub bore and release lever contact face with a little molybdenum disulphide grease.

7 Clutch cable - removal and refitting

1 Jack-up the front of the car and support securely under the front crossmember.
2 Release the locknut on the outer cable at the bellhousing and back the adjusting nut right off.
3 Pull back the rubber gaiter on the end of the release arm and unhook the cable end.
4 Where applicable remove the cowl trim from the instrument panel (6 self-tapping screws).
5 Disconnect the clutch cable from the pedal by pushing the pin out of the cable eye. The clutch cable can now be withdrawn.
6 Refitting the new cable is a reverse of the removal procedure, following which it will be necessary to adjust the pedal free-travel as described in Section 2. Apply a little general purpose grease to the cable end-fitting and clutch pedal pivot.

8 Clutch pedal - removal and refitting

1 Remove the cowl trim from the instrument panel (6 self-tapping screws).
2 Carefully prise off the left-hand spring clip and washer from the pedal pivot shaft.
3 Unhook the clutch pedal return spring.
4 Push the pedal shaft to the side to permit the pedal to be removed.
5 Disconnect the clutch cable from the pedal by pushing the pin out of the eye.
6 Remove the clutch pedal pivot bush.
7 Refitting is the reverse of the removal procedure, but a new spring clip should be used on the pedal shaft for safety's sake, and a little general purpose grease applied to the pedal pivot. Also, if there is wear in the pivot bush, a replacement should be fitted. On completion, adjust the pedal free-travel, as described in Section 2.

9 Fault diagnosis - clutch

There are four main faults to which the clutch and release mechanism are prone. They may occur by themselves, or in conjunction with any of the other faults. They are: clutch squeal, slip, spin and judder.

Clutch squeal - diagnosis and remedy

1 If, on taking up the drive or when changing gear, the clutch squeals this is indicative of a badly worn clutch release bearing.
2 As well as regular wear due to normal use, wear of the clutch release bearing is much accentuated if the clutch is ridden or held down for long periods in gear, with the engine running. To minimise wear of this component the car should always be taken out of gear at traffic lights and for similar hold ups.
3 The clutch release bearing is not an expensive item, but is difficult to get at.

Clutch slip - diagnosis and remedy

1 Clutch slip is a self-evident condition which occurs when the clutch disc is badly worn, oil or grease have got onto the flywheel or pressure plate faces, or the pressure plate itself is faulty.
2 The reason for clutch slip is that, due to one of the faults listed above, there is either insufficient pressure from the pressure plate, or insufficient friction from the clutch disc to ensure a positive drive.
3 If small amounts of oil get onto the clutch, they will be burnt off

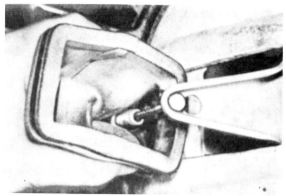

Fig. 5.5. Removing the release arm rubber gaiter, to unhook the clutch cable (Sec. 7)

under the heat of the clutch engagement, and in the process, gradually darken the linings. Excessive oil on the clutch will burn off leaving a carbon deposit which can cause quite bad slip, or fierceness, spin and judder.
4 If clutch slip is suspected, and confirmation of this condition is required, there are several tests which can be made.
5 With the engine in second or third gear and pulling lightly up a moderate incline, sudden depression of the accelerator pedal may cause the engine to increase its speed without any increase in road speed. Easing off from the accelerator will then give a definite drop in engine speed without the car slowing.
6 In extreme cases of clutch slip the engine will race under normal acceleration conditions.

Clutch spin

1 Clutch spin, is a condition which occurs when the release arm travel is excessive, there is an obstruction in the clutch, either on the primary gear splines, or in the operating lever itself, or the oil may have partially burnt off the clutch linings and have left a resinous deposit which is causing the clutch disc to stick to the pressure plate or flywheel.
2 The reason for clutch spin is that due to any, or a combination, of the faults just listed, the clutch pressure plate is not completely freeing from the disc even with the clutch pedal fully depressed.
3 If clutch spin is suspected, the condition can be confirmed by extreme difficulty in engaging first gear from rest, difficulty in changing gear, and very sudden take-up of the clutch drive at the fully depressed end of the clutch pedal travel as the clutch is released.
4 Check that the clutch cable is correctly adjusted and if in order, then the fault lies internally in the clutch. It will then be necessary to remove the clutch for examination and to check the gearbox input shaft.

Clutch judder

1 Clutch judder is a self-evident condition which occurs when the gearbox or engine mountings are loose or too flexible, when there is oil on the faces of the clutch disc, or when the clutch pressure plate has been incorrectly adjusted during assembly.
2 The reason for clutch judder is that due to one of the faults just listed, the clutch pressure plate is not freeing smoothly from the disc, and is snatching.
3 Clutch judder normally occurs when the clutch pedal is released in first or reverse gears, and the whole car shudders as it moves backward or forward.

Chapter 6 Gearbox

For modifications, and information applicable to later models, see Supplement at end of manual

Contents

Fault diagnosis - gearbox ... 8	Gearbox - removal and refitting ... 2
Gearbox - dismantling ... 3	General description ... 1
Gearbox - examination and renovation ... 4	Input shaft - dismantling and reassembly ... 5
Gearbox - reassembly ... 7	Mainshaft - dismantling and reassembly ... 6

Specifications

Number of gears ...	4 forward, 1 reverse
Type of gears ...	Helical, constant mesh
Gear ratios:	
First ...	3.58 to 1
Second ...	2.01 to 1
Third ...	1.397 to 1
Top ...	1.00 to 1
Reverse ...	3.324 to 1
Lubricant type ...	SAE 80 EP gear oil
Oil capacity ...	1 litre (1.75 pints)
Laygear cluster endfloat ...	0.006 to 0.018 in (0.15 to 0.45 mm)
Layshaft diameter ...	0.68 in (17.3 mm)
Thrust washer thickness ...	0.061 to 0.063 in (1.55 to 1.60 mm)

Torque wrench settings	lb f ft	kg f m
Clutch bellhousing to transmission ...	31 to 35	4.2 to 4.8
Clutch bellhousing to engine:		
2 top bolts ...	29 to 35	3.9 to 4.8
Other bolts ...	22 to 27	3.0 to 3.7
Input shaft retainer bolts ...	15 to 18	2.1 to 2.5
Extension housing retaining bolts ...	33 to 36	4.5 to 4.9
Transmission cover bolts ...	15 to 18	2.1 to 2.5

1 General description

The manually operated gearbox is equipped with four forward gears and one reverse.

All forward gears are engaged through blocker ring synchromesh units to obtain smooth, silent gearchanges. All forward gears on the mainshaft and input shaft are in constant mesh with their corresponding gears on the layshaft gear cluster and are helically cut to achieve quiet running.

The layshaft reverse gear has straight-cut spur teeth and drives the toothed 1st/2nd gear synchronizer hub on the mainshaft through an interposed sliding idler gear.

Gears are engaged by a single selector rod and forks and control of the gears is from a floor mounted shift lever.

Where close tolerances and limits are required during assembly of the gearbox, selective shims are used to eliminate excessive endfloat or backlash. This eliminates the need for using matched assemblies.

2 Gearbox - removal and refitting

1 The gearbox can be removed either in unit with the engine or separately from underneath the car. If it is to be removed with the engine, first disconnect the engine and its attachments as described in Chapter 1, Section 5, paragraphs 1 to 20 and then paragraphs 5 to 12 of this Section. The engine and gearbox can then be removed from the car. An assistant will be required to help guide the engine and gearbox through the surrounding body components and fittings.

2 If the gearbox alone is to be removed from the car, it can be taken out from below leaving the engine in position. It will mean that a considerable amount of working room is required beneath the car, and ideally ramps or an inspection pit should be used. However, provided that suitable jacks and supports are available, the task can be accomplished without the need for sophisticated equipment. A cranked spanner may be required to enable the gearshift lever to be removed.

3 Disconnect the battery earth lead.

4 Disconnect the starter motor leads and remove the starter motor (two bolts - refer to Chapter 10 if necessary).

5 If a parcel tray or centre console are fitted, remove these items (refer to Chapter 12, if necessary).

6 Remove the gearlever gaiter(s), bend back the lock tab on the retainer and unscrew the retainer using a suitably cranked spanner. Lift out the gearlever (photos).

7 Remove the propeller shaft, as described in Chapter 7. A polythene bag must be tied around the end of the gearbox to prevent loss of oil.

8 Pull back the rubber gaiter from the clutch release lever and slacken the clutch cable adjuster so that the cable can be unhooked from the release lever.

9 Remove the leads from the reverse light switch, noting which way they are fitted (photo).

10 Remove the circlip retaining the speedometer drive on the gearbox extension housing.

11 Temporarily take the gearbox weight with a trolley jack. Detach the gearbox mounting from the body (4 bolts). If wished, the mounting

2.6a The gaiter used where there is no console

2.6b The rubber gaiter

Fig. 6.1. Remove the gearshift lever (Sec. 2)

2.6c Removing the gear lever

2.9 The reverse light lead cover/connection

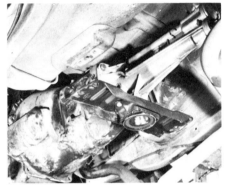
2.11 Detach the gearbox mounting

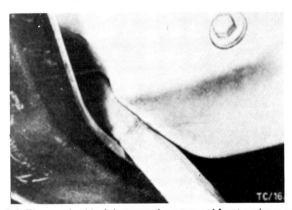
Fig. 6.2. The wooden block between the sump and front engine mounting (Sec. 2)

Fig. 6.3. Clutch housing guide bushes (arrowed) (Sec. 2)

may also be removed from the gearbox (photo).
12 Remove the single adaptor plate bolt and the six bolts at the gearbox flange.
13 Lift the gearbox away from the engine a little then turn it through 90°. Insert a wooden block between the sump and the front engine mounting to prevent the engine from dropping, then withdraw the gearbox rearwards and downwards.
14 When refitting, ensure that the two clutch housing guide bushes are fitted to the engine and that the clutch pilot bearing in the flywheel is fitted and is serviceable. Tie the clutch release lever to the clutch housing with a piece of wire or string to prevent the release lever from slipping out.
15 Apply a little general purpose grease to the gearbox input shaft splines, then install the gearbox using the reverse procedure to that used when removing it.
16 Prior to removing the car from the inspection pit, ramps or jacks, refill the gearbox with the correct grade and quantity of oil, and adjust the clutch free-play as described in Chapter 5.

3 Gearbox - dismantling

1 Remove the clutch release bearing from the gearbox input shaft (photo).
2 Then lift out the clutch release lever (photo).
3 Undo and remove the four bolts holding the bellhousing to the gearbox (photo).
4 Detach the bellhousing from the gearbox (photo).
5 Place the gearbox on a suitable workbench with blocks available to use as supports while dismantling.
6 Referring to Fig. 6.4 undo the four bolts holding the gearbox top cover (1) in place (photo A) and remove the cover (photo B).
7 Remove the spring (7) and detent ball (8). The ball can either be removed using a magnet or a screwdriver with a blob of grease on the end.
8 Tip the gearbox over onto one side and drain the oil into a suitably sized container.
9 Prise out the cup shaped speedometer drive cover (31) on the side

Chapter 6/Gearbox

3.1 Remove the clutch release bearing

3.2 Lift out the release lever

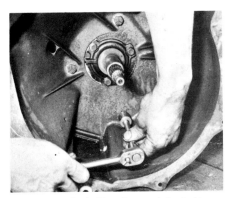
3.3 Remove the bellhousing retaining bolts

3.4 Detach the bellhousing

3.6a Undo the bolts ...

3.6b ... and remove the top cover

3.9 Prise out the cap shaped retainer plug

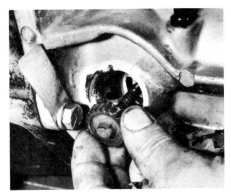
3.10 Remove the speedometer gear

3.11 Drive out the rear extension cover

of the gearbox extension (photo).
10 From under this seal pull out the speedometer gear (30) (photo). To remove it, it may be necessary to tap it from the other end.
11 From where the gearlever enters the extension housing, drive out the rear extension cover (23) (photo).
12 Using a small punch drive out the pin holding the selector boss to the central rod (photo).
13 Now withdraw the selector rod (photo A) rearwards, at the same time holding the selector boss and lock plate (photo B) to prevent them from falling into the gearbox.
14 To remove the selector forks, it is now necessary to push or drive the two synchro hubs towards the front of the gearbox. This can be done with a small punch or a screwdriver; now lift out the selector forks.
15 Turn now to the gearbox extension (21) and remove the bolts and washers which hold it to the gearbox casing.
16 Knock it slightly rearwards with a soft headed hammer and then rotate the whole extension until the cut-out on the extension face coincides with the rear end of the layshaft in the lower half of the gearbox casing.
17 Manufacture a metal rod to act as a dummy layshaft 6 13/16 in (173 mm) long with a diameter of 0.68 in (17.3 mm).
18 Tap the layshaft rearwards with a drift until it is just clear of the front of the gearbox casing, then insert the dummy shaft and drive the layshaft out and allow the laygear cluster to drop out of mesh with the mainshaft gears and into the bottom of the casing.
19 Withdraw the mainshaft and extension assembly from the gearbox casing, pushing the 3rd/top synchronizer hub forward slightly to obtain the necessary clearance. A small roller bearing should come away on the nose of the mainshaft, but if it is not there it will be found in its recess in the input shaft and should be removed and placed in a safe place if it is to be reused.
20 Moving to the front of the gearbox, remove the bolts retaining the input shaft bearing retainer (26) and take it off the shaft.

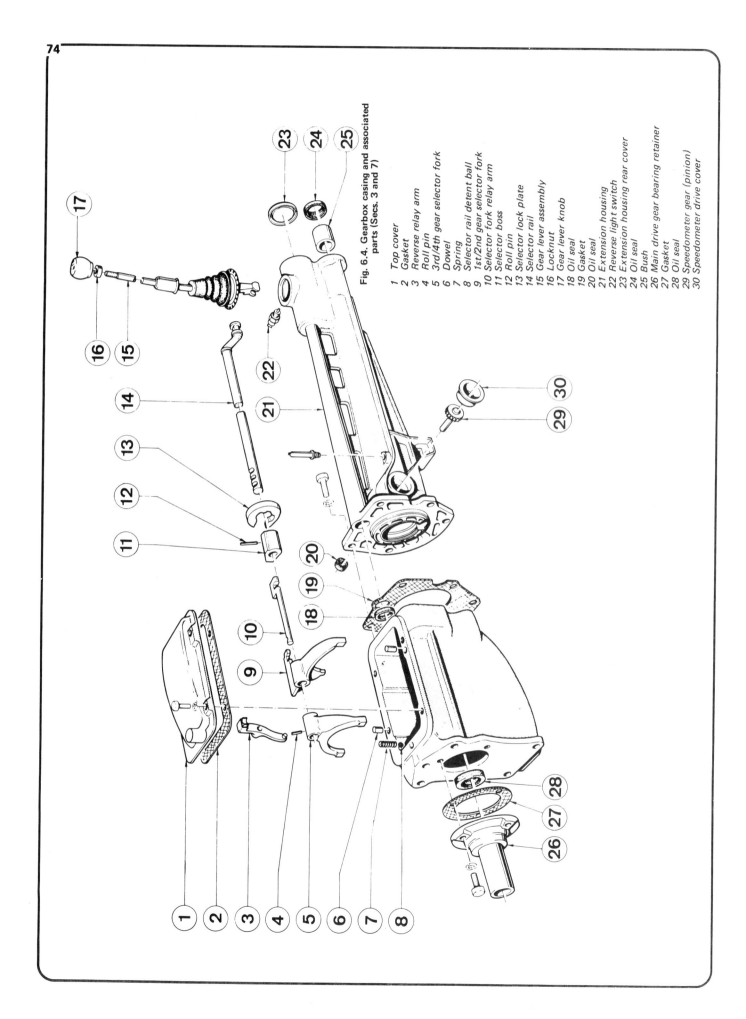

Fig. 6.4. Gearbox casing and associated parts (Secs. 3 and 7)

1 Top cover
2 Gasket
3 Reverse relay arm
4 Roll pin
5 3rd/4th gear selector fork
6 Dowel
7 Spring
8 Selector rail detent ball
9 1st/2nd gear selector fork
10 Selector fork relay arm
11 Selector boss
12 Roll pin
13 Selector lock plate
14 Selector rail
15 Gear lever assembly
16 Locknut
17 Gear lever knob
18 Oil seal
19 Gasket
20 Oil seal
21 Extension housing
22 Reverse light switch
23 Extension housing rear cover
24 Oil seal
25 Bush
26 Main drive gear bearing retainer
27 Gasket
28 Oil seal
29 Speedometer gear (pinion)
30 Speedometer drive cover

Chapter 6/Gearbox

3.12 Drive out the selector boss pin

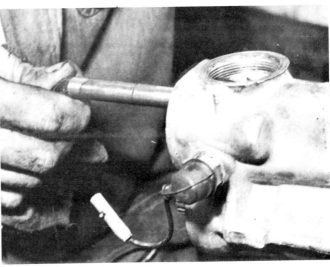

3.13a Withdraw the selector rod ...

3.13b ... whilst holding the selector boss and lock plate

3.24 Remove the reverse idler shaft

21 Remove the large circlip now exposed and then tap, on the bearing outer race, to remove it, and the input shaft, from inside the gearbox.
22 The laygear can now be withdrawn from the rear of the gearbox together with its thrust washers (one at either end).
23 Remove the mainshaft assembly from the gearbox extension, by taking out the large circlip adjacent to the mainshaft bearing (12), Fig. 6.5, then tapping the rear of the shaft with a soft headed hammer. Do not discard this circlip at this stage as it is required for setting-up during reassembly.
24 The reverse idler gear can be removed by screwing a suitable bolt into the end of the shaft and then levering the shaft out with the aid of two large open ended spanners (photo).
25 The gearbox is now completely stripped and must be thoroughly cleaned. If there are any metal chips and fragments in the bottom of the gearbox casing it is obvious that several items will be found to be badly worn. The component parts of the gearbox and laygear should be examined for wear. The input shaft and mainshaft assemblies should be broken down further as described in the following Sections.

4 Gearbox - examination and renovation

1 Carefully clean and then examine all the component parts for general wear, distortion, slackness of fit, and damage to machined faces and threads.

2 Examine the gearwheels for excessive wear and chipping of the teeth. Renew them as necessary.
3 Examine the layshaft for signs of wear, especially the areas upon which the laygear needle roller bearings operate. If a small ridge can be felt at either end of the shaft it will be necessary to renew it.
4 The four synchroniser rings are bound to be worn and it is false economy not to renew them. New rings will improve the smoothness and speed of the gearchange considerably.
5 The needle roller bearing (9) (Fig. 6.5) located between the nose of the mainshaft and the annulus in the rear of the input shaft is also liable to wear, and should be renewed as a matter of course.
6 Examine the condition of the two ball bearing assemblies, one on the input shaft (7) and one on the mainshaft (12). Check them for noisy operation, looseness between the inner and outer races, and for general wear. Normally they should be renewed on a gearbox that is being rebuilt.
7 If either of the synchroniser units (2 and 5) are worn it will be necessary to buy a complete assembly as the parts are not sold individually.
8 Examine the ends of the selector forks where they rub against the channels in the periphery of the synchroniser units. If possible compare the selector forks with new units to help determine the wear that has occurred. Renew them if worn.
9 If the bush bearing in the extension housing is badly worn it is best to take the extension to your local Ford garage to have the bearing

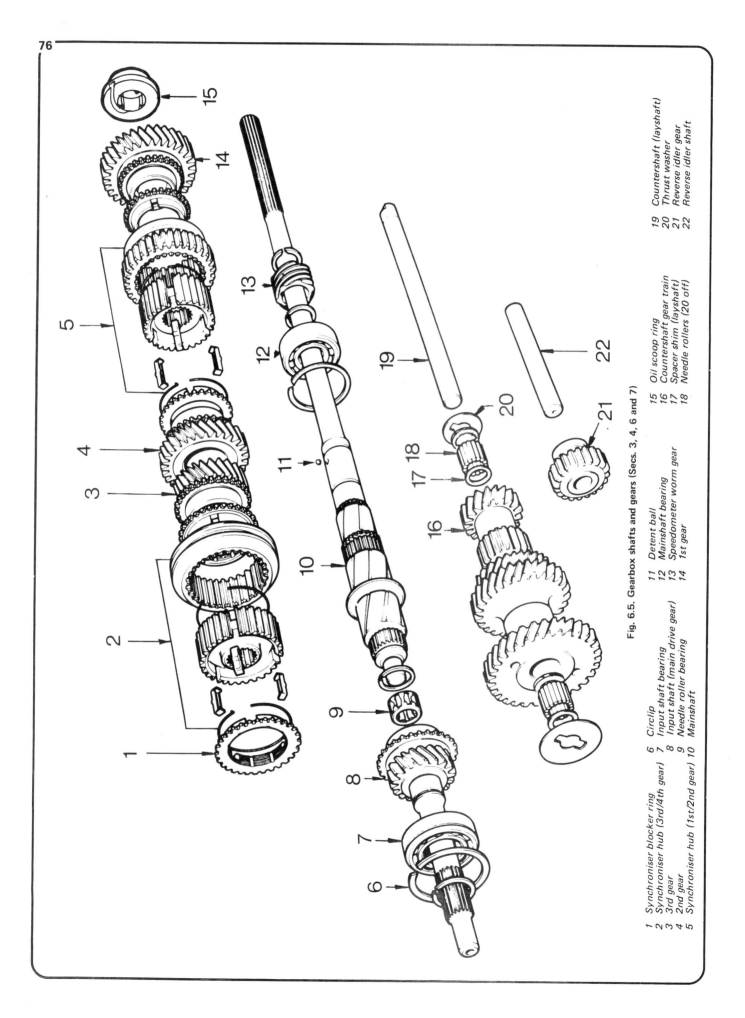

Fig. 6.5. Gearbox shafts and gears (Secs. 3, 4, 6 and 7)

1 Synchroniser blocker ring
2 Synchroniser hub (1st/2nd gear)
3 3rd gear
4 2nd gear
5 Synchroniser hub (3rd/4th gear)
6 Circlip
7 Input shaft bearing
8 Input shaft (main drive gear)
9 Needle roller bearing
10 Mainshaft
11 Detent ball
12 Mainshaft bearing
13 Speedometer worm gear
14 1st gear
15 Oil scoop ring
16 Countershaft gear train
17 Spacer shim (layshaft)
18 Needle rollers (20 off)
19 Countershaft (layshaft)
20 Thrust washer
21 Reverse idler gear
22 Reverse idler shaft

Chapter 6/Gearbox

pulled out and a new one fitted.

10 The oil seals in the extension housing and input shaft bearing retainer should be renewed as a matter of course. Drive out the old seal with the aid of a drift or broad screwdriver. It will be found that the seal comes out quite easily.

11 With a piece of wood to spread the load evenly, carefully tap the new seals into place ensuring that they enter the bore squarely.

12 The only point on the mainshaft that is likely to be worn is the nose where it enters the input shaft. However, examine it thoroughly for any signs of scoring, picking up or flats, and if damage is apparent, renew it.

5 Input shaft - dismantling and reassembly

1 The only reason for dismantling the input shaft is to fit a new ball bearing assembly, or, if the input shaft is being renewed and the old bearing is in excellent condition, then the fitting of a new shaft to an old bearing.

2 With a pair of expanding circlip pliers remove the small circlip which secures the bearing to the input shaft.

3 With a soft headed hammer gently tap the bearing forward and then remove it from the shaft.

4 When fitting the new bearing ensure that the groove cut in the outer periphery faces away from the gear. If the bearing is fitted the wrong way round it will not be possible to fit the large circlip which retains the bearing in the housing.

5 Using the jaws of a vice as a support behind the bearing, tap the bearing squarely into place by hitting the rear of the input shaft with a soft faced hammer.

6 Finally, refit the circlip which holds the bearing to the input shaft.

6 Mainshaft - dismantling and reassembly

1 The mainshaft has to be dismantled before some of the synchroniser rings can be inspected. For dismantling it is best to clamp the plain portion of the shaft between two pieces of wood in a vice.

2 From the forward end of the mainshaft pull off the caged roller bearing (9) and the synchro ring (1) (Fig. 6.5).

3 With a pair of circlip pliers remove the circlip which holds the third/fourth gear synchroniser hub in place.

4 Ease the hub (2) and third gear (3) forward by gentle leverage with a pair of long nosed pliers.

5 The hub (2) and synchro ring are then removed from the mainshaft.

6 Then slide off third gear. Nothing else can be removed from this end of the mainshaft because of the raised lip on the shaft.

7 Move to the other end of the mainshaft and remove the small circlip. Then slide off the speedometer drive (13), taking care not to lose the ball which locates in a groove in the gear and a small recess in the mainshaft.

8 Remove the circlip and then gently lever off the mainshaft large bearing (12) with the aid of two tyre levers as shown in the photo.

9 The bearing, followed by the oil scoop ring (15) can then be pulled off. Follow these items by pulling off first gear (14) and the synchroniser ring.

10 With a pair of circlip pliers remove the circlip which retains the first and second gear synchroniser assembly in place.

11 The first and second gear synchroniser followed by second gear (4) are then slid off the mainshaft. The mainshaft is now completely dismantled.

12 If a new synchroniser assembly is being fitted it is necessary to take it to pieces first to clean off all the preservative. These instructions are

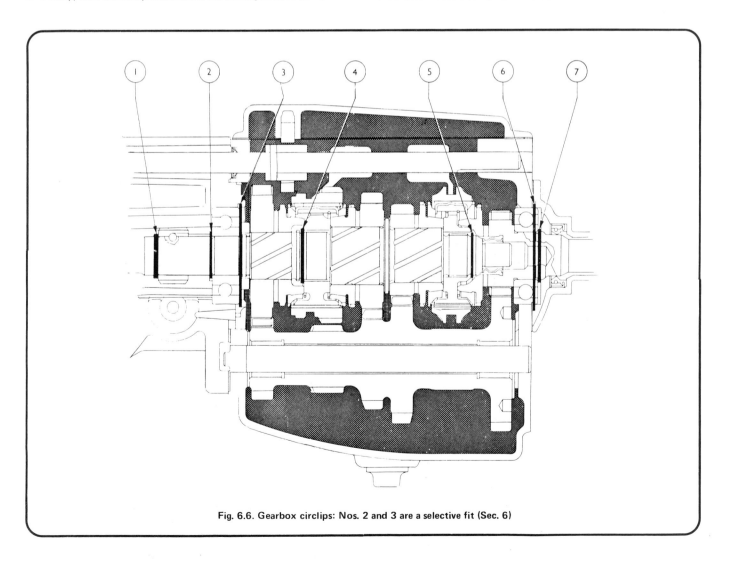

Fig. 6.6. Gearbox circlips: Nos. 2 and 3 are a selective fit (Sec. 6)

6.8 Remove the large bearing

Fig. 6.7. The synchro hub component parts (Sec. 6)

Fig. 6.8. Relative positions of synchro spring clips (Sec. 6)

One spring must be put on anti-clockwise and one clockwise when viewed from the side (see Fig. 6.8). When either side of the assembly is viewed face on, the direction of rotation of the springs should then appear the same.

16 Prior to reassembling the mainshaft read paragraphs 22 and 24 to ensure that the correct thickness of selective circlips can be obtained. Reassembly commences by refitting second gear (4) (Fig. 6.5), gear teeth facing the raised lip, and its synchroniser ring on the rear portion of the mainshaft (photo).

17 Next slide on the first and second gear synchroniser assembly (5) (photo A) and ensure that the cut-outs in the synchroniser ring fit over the blocker bars in the synchroniser hub (photo B); that the marks on the mainshaft and hub are in line (where made), and that the reverse gear teeth cut on the synchroniser sleeve periphery are adjacent to second gear.

18 Refit the circlip which holds the synchroniser hub in place (photo).
19 Then fit another synchroniser ring, again ensuring that the cut-outs in the ring fit over the blocker bars in the synchroniser hub.
20 Next slide on first gear (14) so that the synchronising cone portion lies inside the synchronising ring just fitted (photo).
21 Fit the oil scoop ring (15), large diameter facing the first gear (photo).
22 If a new mainshaft bearing (12) or a new gearbox extension is being used it will now be necessary to select a new large circlip to eliminate endfloat of the mainshaft. To do this, first fit the original circlip in its groove in the gearbox extension and draw it outwards (ie. away from the rear of the extension). Now accurately measure the dimension from the base of the bearing housing to the outer edge of the circlip and record the figure. Also accurately measure the thickness of the bearing outer track (Fig. 6.9) and subtract this figure from the depth already recorded. This will give the required circlip thickness to give zero endfloat.
23 Loosely fit the selected circlip, lubricate the bearing contact surfaces then press it onto the shaft. To press the bearing home, close the jaws of a vice until they are not quite touching the mainshaft and with the bearing resting squarely against the side of the vice jaws, press the bearing on by tapping the end of the shaft with a soft headed hammer (photo).
24 Refit the small circlip retaining the main bearing in place. This is also a selective circlip and must be such that all endfloat between the bearing inner track and the circlip edge is eliminated (photo).
25 Refit the small ball (11) (Fig. 6.5) that retains the speedometer drive, in its recess in the mainshaft (photo).
26 Slide on the speedometer drive (photo), noting that it can only be fitted one way round as the groove in which the ball fits does not run the whole length of the drive.
27 Now fit the circlip to retain the speedometer drive (photo). Assembly of this end of the mainshaft is now complete.
28 Moving to the short end of the mainshaft slide on third gear (3) so that the machined gear teeth lie adjacent to second gear, then slide on the synchroniser ring (photo).
29 Fit the third and fourth gear synchroniser assembly (2) (photo), again ensuring that the cut-outs on the ring line up with the blocker bars.
30 With a suitable piece of metal tube over the mainshaft, tap the synchroniser fully home onto the mainshaft (photo).
31 Then fit the securing circlip in place (photo). Apart from the needle roller bearing which fits on the nose of the mainshaft, this completes mainshaft reassembly.

7 Gearbox - reassembly

1 If removed, refit the reverse idler gear and selector lever in the gearbox, by tapping in the shaft (22) (Fig. 6.5). Once it is through the casing fit the gear wheel (21) so that its gear teeth are facing in towards the main gearbox area.
2 Fit the reverse selector lever in the groove in the idler gear then drive the shaft home with a soft headed hammer until it is flush with the gearbox casing.
3 Slide a spacer shim (17) into either end of the laygear (16), so that they abut the internal machined shoulders.
4 Smear thick grease on the laygear roller bearing surface and fit the needle rollers (18) one at a time (photo), until all are in place. The grease will hold the rollers in position. Build up the needle roller bearings in the other end of the laygear in a similar fashion. Note that

also pertinent in instances where the outer sleeve has come off the hub accidentally before dismantling.
13 To dismantle an assembly for cleaning slide the synchroniser sleeve off the splined hub and clean all the preservative from the blocker bars, spring rings, the hub itself and the sleeve.
14 Oil the components lightly and then fit the sleeve to the hub. Note the three slots in the hub and fit a blocker bar in each.
15 Fit the two springs, one on the front and one on the rear face of the inside of the synchroniser sleeve under the blocker bars with the tagged end of each spring locating in the 'U' section of the same bar.

6.16 Refit the second gear and synchro ring

6.17a Slide on the first and second gear synchroniser assembly ...

6.17b ... and make sure that the ring fits over the blocker bars

6.18 Refit the circlip

6.20 Slide on first gear

6.21 Fit the oil scoop ring

Fig. 6.9. Measuring the mainshaft bearing track (Sec. 6)

6.23 Press the bearing home

6.24 Refit the main bearing selective circlip

6.25 Refit the speedometer drive retaining ball

6.26 Slide on the speedometer drive gear

6.27 Fit the circlip

6.28 Slide on the third gear and synchroniser ring

6.29 Fit the third and fourth gear synchroniser assembly

6.30 Tap the synchroniser fully home

6.31 Fit the securing circlip

7.4 Fit the laygear needle rollers

7.5 Fit the external washers

7.6 Slide in the dummy shaft

7.8 Fit the smaller thrust washer at the rear

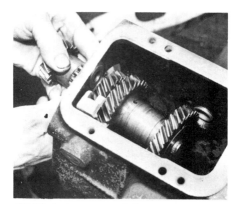
7.9 Fit the lay gear carefully

7.10a Slide in the input shaft assembly ...

7.10b ... and drive the bearing into place

Chapter 6/Gearbox

there should be twenty at each end.

5 Fit the external washer to each end of the laygear, taking care not to dislodge the roller bearings (photo).

6 Carefully slide in the dummy shaft used previously for driving out the layshaft (photo).

7 Grease the two thrust washers (20) and position the larger of the two in the front of the gearbox so that the tongues fit into the machined recesses.

8 Fit the smaller of the thrust washers to the rear of the gearbox in the same way (photo).

9 Fit the laygear complete with dummy shaft in the bottom of the gearbox casing taking care not to dislodge the thrust washers (photo).

10 Now from inside the gearbox, slide in the input shaft assembly (8) (photo A) and drive the bearing into place with a suitable drift (photo B).

11 Secure the bearing in position by refitting the circlip (6) (photo).

12 Fit a new gasket to the input shaft bearing retainer and smear on some non-setting jointing compound (photo).

13 Refit the retainer over the input shaft (photo A) ensuring that the oil drain hole is towards the bottom of the gearbox, and tighten down the bolts (photo B).

14 Submerge the gearbox end of the extension housing in hot water for a few minutes, then mount it in a vice and slide in the mainshaft assembly (photo). Take care that the splines do not damage the oil seal.

15 Secure the mainshaft to the gearbox extension by locating the circlip already placed loosely behind the main bearing into its groove in the extension (photo A). Photo B shows the circlip correctly located.

16 Fit a new gasket to the extension housing and then refit the small roller bearing on the nose of the mainshaft (photo). Lubricate the roller bearing with gearbox oil.

17 Slide the combined mainshaft and extension housing assembly into the rear of the gearbox (photo) and mate up the nose of the mainshaft with the rear of the input shaft.

18 Completely invert the gearbox so that the laygear falls into mesh with the mainshaft gears.

19 Turn the extension housing round until the cut-out on it coincides with the hole for the layshaft (photo). It may be necessary to trim the gasket.

7.11 Fit the circlip to retain the bearing

7.12 Apply sealing compound to the input shaft bearing retainer

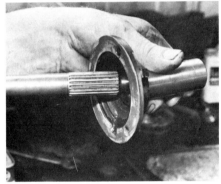

7.13a Refit the input shaft bearing retainer ...

7.13b ... and tighten the bolts

7.14 Slide in the mainshaft assembly

7.15a Locate the circlip in the groove in the extension housing

7.15b The circlip correctly located

7.16 Refit the mainshaft roller bearing

7.17 Slide the mainshaft and extension housing into the gearbox

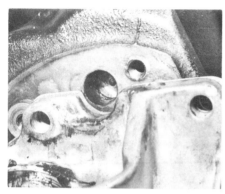
7.19 Turn the extension housing to align the cut out with the hole for the layshaft

7.20 Push in the layshaft

7.21 Ensure that the cut out in the end of the layshaft is in the horizontal position

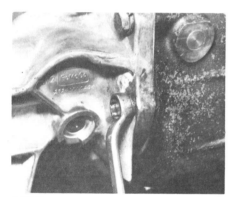
7.22 Secure the extension housing to the gearbox

7.23a Push the 1st/2nd synchroniser hub ...

7.23b ... and the 3rd/4th synchroniser hub fully forward

7.24a Position the selector forks ready ...

7.24b ... to lower in position

20 Push the layshaft into its hole from the rear, thereby driving out the dummy shaft at the same time (photo).

21 Tap the layshaft into position until its front end is flush with the gearbox casing and ensure that the cut-out on the rear end is in the horizontal position so it will fit into its recess in the extension housing flange (photo).

22 Turn the gearbox the right way up again; correctly line up the extension housing and secure it to the gearbox (photo). Apply a non-setting jointing compound to the bolt threads before fitting them.

23 The selector forks cannot be replaced until the two synchroniser hubs are pushed by means of a screwdriver to their most forward positions (photos A and B).

24 Now lower the selector forks into position (photo A); it will be found that they will now drop in quite easily (photo B). Now return the synchroniser hubs to their original positions.

25 Slide the gearchange selector rail into place from the rear of the extension and as it comes into the gearbox housing slide onto it selector boss and lock plate, having made sure that the plate locates in the cut-outs in the selector fork extension arms.

26 Push the selector rod through the boss and the selector forks until the pin holes on the boss and rod align. Tap the pin into place thereby securing the boss to the selector rod. During this operation ensure that the cut-out on the gearbox end of the selector rail faces to the right.

27 Apply a small amount of non-setting jointing compound to the blanking plug and gently tap it into position in the rear of the extension housing behind the selector rod. Peen it with a centre punch in three or four places to retain it.

28 Insert the detent ball and spring (8 and 7 in Fig. 6.4).

29 Place a new gasket on the gearbox top cover plate, then refit the top cover and tighten down its four retaining bolts.

30 Refit the speedometer drive gear in the extension, smear the edges of its retaining cover with non-setting jointing compound and tap the cover into place.

31 Refit the bellhousing onto the gearbox, apply a non-setting jointing compound to the bolt threads before fitting them.

32 Refit the clutch release fork and bearing.

33 The gearbox is now ready for refitting to the car.

8 Fault diagnosis - gearbox

Symptom	Reason/s	Remedy
Weak or ineffective synchromesh	Synchronising cones worn, split or damaged	Dismantle and overhaul gearbox. Fit new gear wheels and synchronising cones.
	Baulk ring synchromesh dogs worn, or damaged	Dismantle and overhaul gearbox. Fit new baulk ring synchromesh.
Jumps out of gear	Broken gearchange fork rod spring	Dismantle and replace spring.
	Gearbox coupling dogs badly worn	Dismantle gearbox. Fit new coupling dogs.
	Selector fork rod groove badly worn	Fit new selector fork rod.
Excessive noise	Incorrect grade of oil in gearbox or oil level too low	Drain, refill or top up gearbox with correct grade of oil.
	Bush or needle roller bearings worn or damaged	Dismantle and overhaul gearbox. Renew bearings.
	Gear teeth excessively worn or damaged	Dismantle, overhaul gearbox. Renew gear wheels.
	Layshaft thrust washers worn allowing excessive end play	Dismantle and overhaul gearbox. Renew thrust washers.
Excessive difficulty in engaging gear	Clutch cable adjustment incorrect	Adjust clutch cable correctly.

Chapter 7 Propeller shaft

Contents

Fault diagnosis - propeller shaft ... 4	Propeller shaft - removal and refitting ... 2
General description ... 1	Universal joints - inspection ... 3

Specifications

Type ... Single section open shaft with two universal joints

Torque wrench setting

	lbf ft	kgf m
Propeller shaft to drive pinion flange ...	44 to 48	6.0 to 6.5

1 General description

On the Capri II, 1300 cc models, the drive is transmitted from the gearbox to the rear axle by a single, finely balanced, tubular propeller shaft.

A universal joint is fitted at each end of the propeller shaft and these allow for the vertical movement of the rear axle and any movement of the power unit on its rubber mountings. Each universal joint comprises a four legged centre spider, four needle roller bearings and two yokes.

The fore and aft movement of the rear axle is absorbed by a sliding spline located at the gearbox end. The yoke flange of the rear universal joint is fitted to the rear axle and is secured to the pinion flange by four bolts and lockwashers.

The propeller shaft universal joints cannot be repaired on a do-it-yourself basis since the joint spiders are located into the yoke in a position determined during electronic balancing. Therefore, when joint wear is detected, either a replacement propeller shaft must be obtained or the complete propeller shaft must be passed to a suitably equipped

Fig. 7.1. The single section propeller shaft unit

2.5 Removing the propeller shaft from the axle

2.7 Disengaging the propeller shaft from the gearbox

Chapter 7/Propeller shaft

engineering workshop for repair.
 No special tools are required for the removal or replacement of the propeller shaft unit.

2 Propeller shaft - removal and refitting

1 Jack-up the rear of the car, or position the rear of the car over a pit or on a ramp.
2 If the rear of the car is jacked-up, supplement the jack with support blocks so that danger is minimised, should the jack collapse.
3 If the rear wheels are off the ground, place the car in gear or put the handbrake on to ensure that the propeller shaft does not turn when an attempt is made to loosen the four bolts securing the propeller shaft to the rear axle companion flange.
4 The propeller shaft is carefully balanced to fine limits and it is important that it is replaced in exactly the same position it was in, prior to its removal. Scratch a mark on the propeller shaft and rear axle flanges to ensure accurate mating when the time comes for refitting.
5 Unscrew and remove the four lock bolts and securing washers which hold the flange on the propeller shaft to the flange on the rear axle (photo).
6 Push the shaft forward slightly to separate the two flanges to the rear, then lower the end of the shaft and pull it rearwards to disengage it from the gearbox mainshaft splines (photo).
7 Place a large can or tray under the rear of the gearbox extension to catch any oil which is likely to leak through the spline lubricating holes when the propeller shaft is removed.
8 Refitting the propeller shaft is the reversal of removal, but ensure that the mating marks scratched on the propeller shaft and axle flanges line up.
9 When fitting the propeller drive shaft into the transmission extension take care not to damage the oil seal in the extension housing.
10 Tighten the rear axle drive pinion to the propeller shaft flange retaining bolts to the specified torque and always fit new lock washers.
11 Check the oil level in the gearbox and replace any that may have leaked through the rear extension during the period the propeller shaft was removed.

3 Universal joints - inspection

1 Wear in the needle roller bearings is characterised by vibration in the transmission, 'clonks' on taking up the drive, and in extreme cases of lack of lubrication, metallic squeaking, and ultimately grating and shrieking sounds as the bearings break up.
2 It is easy to check if the needle roller bearings are worn with the propeller shaft in position, by trying to turn the shaft with one hand and holding the rear axle flange with the other, when the rear universal is being checked. Check the front universal by holding the front coupling in one hand and the prop in the other and turning in opposite directions. Any movement between the propeller shaft and the front and the rear half couplings is indicative of considerable wear. If worn a replacement propeller shaft must be obtained, or the existing shaft overhauled by a suitably equipped engineering workshop.

4 Fault diagnosis - propeller shaft

Symptom	Reason/s
Vibration	Wear in sliding splines. Worn universal joint bearings. Propeller shaft out of balance. Distorted propeller shaft.
Knock or 'clunk' when taking up drive	Worn universal joint bearings. Worn rear axle drive pinion splines. Loose rear drive flange bolts. Excessive backlash in rear axle gears. Roadwheel retaining nuts loose

Chapter 8 Rear axle

Contents

Differential - overhaul ... 6	Halfshaft - removal and refitting ... 3
Differential carrier - removal and refitting ... 5	Halfshaft bearing/oil seal - removal and refitting ... 4
Fault diagnosis - rear axle ... 8	Pinion oil seal - removal and refitting ... 7
General description ... 1	Rear axle - removal and refitting ... 2

Specifications

Axle designation ...	'J' (Timken)
Type ...	Hypoid semi-floating detachable carrier differential
Ratio ...	4.125 : 1
Lubricant capacity ...	2 Imp pints (1.14 litres - 2.4 US pints)
Lubricant type ...	SAE 90 EP oil
Backlash ...	0.005 to 0.007 in (0.127 to 0.178 mm)
Collapsible spacer width ...	1.917 to 1.925 in (48.7 to 48.9 mm)
Differential case bearing pre-load ...	0.008 to 0.010 in (0.20 to 0.25 mm)

Torque wrench settings	lbf ft	kgf m
Bearing cap to axle casing	46 to 51	6.2 to 6.9
Differential assembly to axle casing	26 to 30	3.5 to 4.1
Crownwheel to differential case	51 to 56	6.9 to 7.6
Drive pinion flange bolts	44 to 48	6.0 to 6.5
Halfshaft retainer plate to axle flange	15 to 18	2.1 to 2.5

1 General description

The rear axle is of the hypoid semi-floating type and is located by semi-elliptic leaf springs in conjunction with a stabilizer bar.

The rear axle - Ford designation 'J' type - has a completely detachable differential carrier.

Special tools and gauges are required to overhaul the rear axle, and it is not therefore recommended that a person with limited facilities should undertake this task. However, for those who have the necessary equipment, the overhaul procedure is described in this Chapter.

It is recommended that the differential unit is either renewed on an exchange basis, or the original unit taken to your dealer for reconditioning.

2 Rear axle - removal and refitting

1 Remove the rear wheel hub caps, if fitted, and loosen the wheel nuts.
2 Raise and support the rear of the body and the differential casing with blocks or jacks so that the rear wheels are clear of the ground. This is most easily done by placing a jack under the centre of the differential, jacking up the axle and then fitting chocks under the mounting points at the front of the rear springs to support the body.
3 Remove both rear wheels and place the wheel nuts in the hub caps for safe keeping.
4 Mark the propeller shaft and differential drive flanges to ensure replacement in the same relative positions. Undo and remove the bolts holding the two flanges together.
5 Release the handbrake and detach the cable from the rear axle (Fig. 8.2).
6 Unscrew the union on the brake pipe at the junction on the rear axle (Fig. 8.3) and have handy either a jar to catch the hydraulic fluid or a plug to block the end of the pipe.
7 Disconnect the stabiliser bar brackets.
8 Undo the self-locking nuts holding the shock absorbers to the spring plates thus freeing their lower ends. It may be necessary to slightly raise the jack under the axle casing to successfully free the shock absorbers.
9 Unscrew the nuts from under the spring retaining plates. These nuts screw onto the ends of the inverted 'U' bolts which retain the axle to the spring. Remove the 'U' bolts, bump stops and spring retaining plates.
10 The axle assembly will now be resting free on the centre jack and can now be removed by lifting it through the space between the road spring and the bodyframe sidemember.
11 Reassembly is a direct reversal of the removal procedure, but note the following:

   a) The 'U' bolt nuts must be tightened down to a torque of 18 to 27 lb f ft (2.5 to 3.6 kg f m), the brakes must be bled, and the handbrake readjusted as described in Chapter 9.
   b) When refitting the propeller shaft to the pinion flange, ensure that the alignment marks correspond and use new lockwashers under the bolt heads. Tighten the bolts to the recommended torque of 44 to 48 lb f ft (6.0 to 6.5 kg f m).
   c) The shock absorber and stabilizer bar retaining nuts must only be tightened when the springs are under load.
   d) If the axle has been drained, top up with oil.

Chapter 8/Rear axle

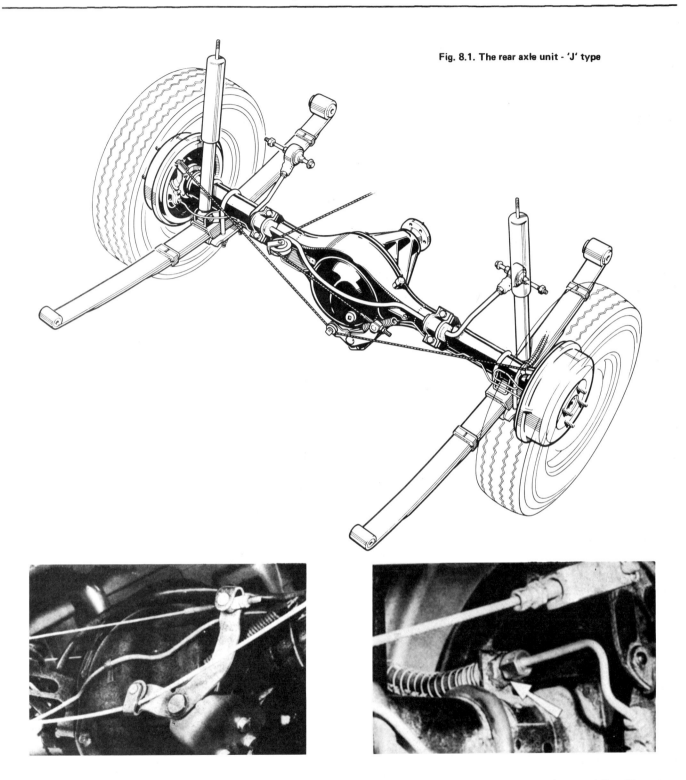

Fig. 8.1. The rear axle unit - 'J' type

Fig. 8.2. The handbrake cable and relay lever attached to the axle (Sec. 2)

Fig. 8.3. The brake hose connection (arrowed) (Sec. 2)

3 Halfshaft - removal and refitting

1 Jack up the rear of the car and support the bodyframe and axle casing with blocks and axle stands. Chock the front wheels.
2 Remove the roadwheel and then unscrew the brake drum retaining screw. Withdraw the brake drum, having first released the handbrake.
3 If the drum is tight, tap the drum from its location using a block of wood or soft faced mallet.
4 Remove the four self-locking nuts retaining the bearing flange plate to the endface of the axle housing. The nuts are accessible through a hole in the halfshaft flange (Fig. 8.4).
5 If available, a slide hammer can be attached to the roadwheel studs and the halfshaft, complete with bearing/seal assembly, extracted from the axle casing (Fig. 8.5).
6 If a slide hammer is not available, it is sometimes possible to extract the halfshafts by attaching an old roadwheel to the hub and then striking two opposite points on the inner wheel rim simultaneously.
   Another method of withdrawing the halfshaft is to use two or

three bolts, fitted with nuts, placed between the halfshaft flange and the axle housing end flange. By evenly unscrewing the nuts fitted to the bolts, the effective length of the bolts will be increased and the halfshaft forced outwards. However, great care must be taken or damage can result.

7 Refitting is simply a matter of inserting the halfshaft into the housing and supporting it horizontally until the splines of the halfshaft can be felt to engage with those of the differential gears. A little grease should be smeared onto the outer surface of the hub bearing in order to prevent future seizure by rust.

4 Halfshaft bearing/oil seal - removal and refitting

1 Withdraw the halfshaft, as described in the preceding Section.
2 Secure the assembly in a vice, the jaws of which have been fitted with soft metal protectors.
3 Drill a hole in the bearing securing collar and then remove the collar by splitting it with a cold chisel. Take care not to damage the shaft during these operations.
4 Using a suitable press, draw off the combined bearing/oil seal.
5 To the halfshaft install the bearing retainer plate, the new bearing (seal side towards the differential) and a new bearing collar. Coat the bearing with a little lithium grease before refitting.
6 Apply pressure to the collar only, using a press or bearing puller, seat the components against the shoulder of the halfshaft flange.
7 Install the halfshaft, as described in the preceding Section.

Fig. 8.4. Removing the bearing retainer plate nuts (Sec. 3)

5 Differential carrier - removal and refitting

1 To remove the differential carrier assembly, drain the oil from the axle by removing the drain plug in the base of the banjo casing, (if

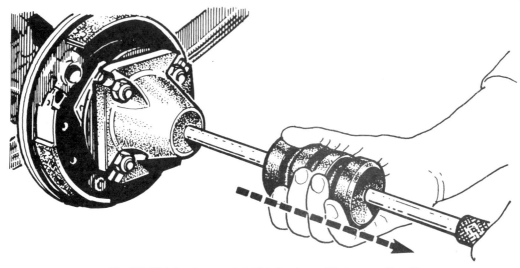

Fig. 8.5. Withdrawing an axle halfshaft using a slide hammer (Sec. 3)

fitted), jack up the rear of the vehicle, remove both roadwheels and brake drums and then partially withdraw both halfshafts as described in Section 3.
2 Disconnect the propeller shaft at the rear end, as described in Chapter 7.
3 Undo the eight self-locking nuts holding the differential carrier assembly to the axle casing. If an oil drain plug has not been fitted, pull the assembly slightly forward and allow the oil to drain in a suitable tray or bowl. The carrier complete with the crownwheel can now be lifted clear with the gasket.
4 Before refitting, carefully clean the mating surfaces of the carrier and the axle casing and always fit a new gasket. Refitting is then a direct reversal of the above instructions. The eight nuts retaining the differential carrier assembly to the axle casing should be tightened to the specified torque. Top up the oil level with the specified grade and quantity of oil.

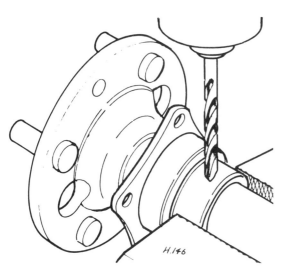

Fig. 8.6. Drilling a hole in the bearing inner ring prior to splitting with a cold chisel (Sec. 4)

6 Differential - overhaul

Most professional garages will prefer to renew the complete differential carrier assembly as a unit if it is worn, rather than dismantle the unit to renew any damaged or worn parts. To do the job correctly 'according to the book' requires the use of special and expensive tools which the majority of garages do not have.

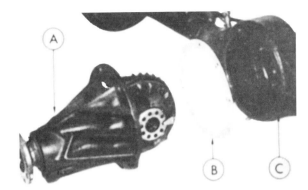

Fig. 8.7. Always fit a new gasket (B), between the differential housing (A), and the axle casing (C) (Sec. 5)

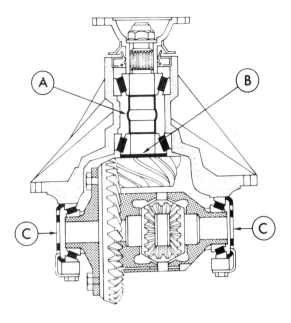

Fig. 8.9. Sectional view of the differential unit (Sec. 6)

A Collapsible spacer   B Shim   C Adjusting nuts

Fig. 8.8. Identification marks on the bearing caps (Sec. 6)

Fig. 8.10. Component parts of the differential (Sec. 6)

The primary object of these special tools is to enable the mesh of the crownwheel to the pinion to be very accurately set and thus ensure that noise is kept to a minimum. If any increase in noise cannot be tolerated (provided that the rear axle is not already noisy due to a defective part) then it is best to purchase an exchange built up differential unit.

The differential assembly should be stripped as follows:

1 Remove the differential assembly from the rear axle, as described in Section 5.
2 With the differential assembly on the bench begin dismantling the unit.
3 Undo and remove the bolts, spring washers and lock plates securing the adjustment cups to the bearing caps.
4 Release the tension on the bearing cap bolts and unscrew the differential bearing adjustment cups. Note from which side each cup originated and mark with a punch or scriber (Fig. 8.8).
5 Unscrew the bearing cap bolts and spring washers. Ensure that the caps are marked so that they may be fitted in their original positions upon reassembly.
6 Pull off the caps and then lever out the differential unit complete with crownwheel and differential gears.
7 Recover the differential bearing outer tracks and inspect the bearings for wear or damage. If evident the bearings will have to be renewed.
8 Using a universal puller and suitable thrust block draw off the old bearings.
9 Undo and remove the bolts and washers that secure the crownwheel to the differential cage. Mark the relative positions of the cage and crownwheel if new parts are not to be fitted and lift off the crownwheel.
10 Clamp the pinion flange in a vice and then undo the nut. Any damage caused to the edge of the flange by the vice should be carefully filed smooth.
11 With the nut removed pull off the splined pinion flange. Tap the end of the pinion shaft with a soft faced mallet, if the flange appears to be stuck.
12 The pinion, complete with spacer and rear bearing cone, may now be extracted from the rear of the housing.
13 Using a drift, carefully tap out the pinion front bearing and oil seal.
14 Check the bearings for signs of wear and if evident the outer tracks must be removed using a suitable soft metal drift.
15 To dismantle the pinion assembly, detach the bearing spacer and remove the rear bearing cone using a universal puller. Recover any shims found between the rear bearing and pinion head.
16 Tap out the differential pinion shaft locking pin which is tapered at one end and must be pushed out from the crownwheel side of the case.
17 Push the differential pinion shaft out of the case and rotate the pinions around the differential gears, so that they may be extracted through the apertures in the case. Cupped thrust washers are fitted between the pinions and the case and may be extracted after the pinions have been removed.
18 Remove the differential gears and thrust washers from the differential case.
19 Wash all parts and wipe dry with a clean lint free cloth.
20 Again check all bearings for signs of wear or pitting and if evident a new set of bearings should be obtained.
21 Examine the teeth of the crownwheel and pinion for pitting, score marks, chipping and general wear. If a crownwheel and pinion is required a mated crownwheel and pinion must be fitted and under no circumstances may only one part of the two be renewed.
22 Inspect the differential pinions and side gears for signs of pitting, score marks, chipping and general wear. Obtain new gears as necessary.
23 Inspect the thrust washers for signs of wear or deep scoring. Obtain new thrust washers as necessary.

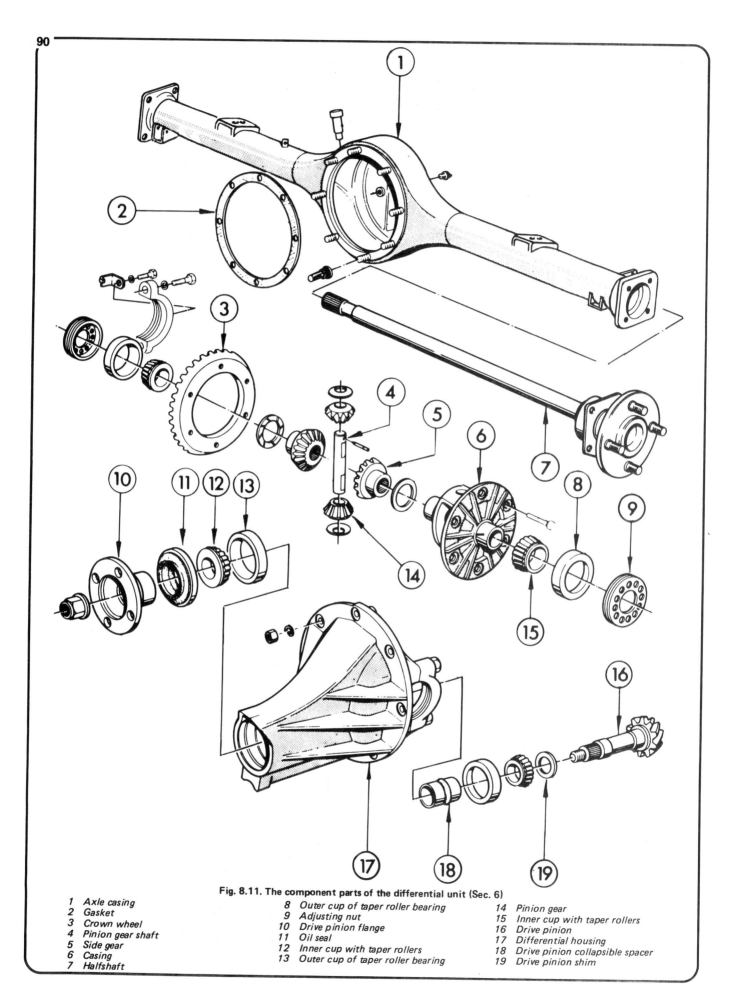

Fig. 8.11. The component parts of the differential unit (Sec. 6)

1 Axle casing
2 Gasket
3 Crown wheel
4 Pinion gear shaft
5 Side gear
6 Casing
7 Halfshaft
8 Outer cup of taper roller bearing
9 Adjusting nut
10 Drive pinion flange
11 Oil seal
12 Inner cup with taper rollers
13 Outer cup of taper roller bearing
14 Pinion gear
15 Inner cup with taper rollers
16 Drive pinion
17 Differential housing
18 Drive pinion collapsible spacer
19 Drive pinion shim

24 Once the pinion oil seal has been disturbed it must be discarded and a new one obtained.

25 Commence reassembly by lubricating the differential gear thrust washers and then positioning a flat washer on each differential side gear. The washers must be fitted with the grooved sides to the side gears. Position the two gears in the case.

26 Position the cupped thrust washers on the machined faces in the case and retain in position with a smear of grease.

27 Locate the pinion gears in the case diametrically opposite each other and rotate the gears to move the pinion gears in line with the holes in the shaft.

28 Check that the thrust washers are still in place and push the spider shaft through the case, thrust washers and pinions. If the pinions do not line up they are not diametrically opposite each other, and should be extracted and repositioned. Measure the play of the gears, and, if necessary select new thrust washers to obtain 0.006 in (0.15 mm) play.

29 Insert the locking pin (tapered end first) and lightly peen the case to prevent the pin working out.

30 Examine the bearing journals on the differential case for burrs, and refit the differential bearing cones onto the differential case using a suitable diameter tubular drift. Make sure they are fitted the correct way round.

31 Examine the crownwheel and differential case for burrs, score marks and dirt. Clean as necessary and then refit the crownwheel. Take care to line up the bolt holes and any previously made marks if the original parts are being refitted.

32 Refit the crownwheel to differential case securing bolts and tighten in a diagonal manner to the specified torque wrench setting.

33 Using a suitable diameter drift carefully drive the pinion bearing cups into position in the final drive housing. Make sure they are the correct way round.

34 Slide the shim onto the pinion shaft and locate behind the pinion head and then fit the inner cone and race of the rear bearing. It is quite satisfactory to drift the rear bearing on with a piece of tubing 12 to 14 inches long with sufficient internal diameter to just fit over the pinion shaft. With one end of the tube bearing against the race, tap the top end of the tube with a hammer so driving the bearing squarely down the shaft and hard up against the underside of the thrust washer.

35 Slide a new collapsible type spacer over the pinion shaft and insert the assembly into the differential carrier.

36 Fit the pinion front bearing outer track and race, followed by a new pinion oil seal.

37 Fit the pinion drive flange and screw on the pinion self-locking nut until a pinion endfloat exists of between 0.002 and 0.005 in (0.05 and 0.13 mm). Tighten the nut only a fraction at a time and check the pinion turning torque after each tightening, using either a suitable torque gauge or a spring balance and length of cord wrapped round the pinion drive flange. The correct pinion turning torque should be:

Original bearings
Torque wrench   12 to 18 lb f in   (0.14 to 0.216 kgfm)
Pull on spring balance 12 to 18 lb   (5 to 8 kg)

New bearings
Torque wrench   20 to 26 lb f in   (0.24 to 0.31 kgfm)
Pull on spring balance 20 to 26 lb   (9 to 11 kg)

38 To the foregoing figures, add 3lb f in (0.035 kg f m) if a new pinion oil seal has been fitted.

39 Throughout the nut tightening process, hold the pinion flange quite still with a suitable tool.

40 If the pinion nut is overtightened, the nut cannot be unscrewed to correct the adjustment as the pinion spacer will have been over-compressed and the assembly will have to be dismantled and a new collapsible type spacer fitted.

41 Fit the differential cage to the differential carrier and refit the two bearing caps, locating them in their original positions.

42 Tighten the bearing cap bolts finger-tight and then screw in the two adjustment cups.

43 It is now necessary to position the crownwheel relative to the pinion. If possible mount a dial indicator gauge, and with the probe resting on one of the teeth of the crownwheel determine the backlash. Backlash may be varied by moving the whole differential assembly using the two adjustment cups until the required setting is obtained.

44 Tighten the bearing cap securing bolts and recheck the backlash

Fig. 8.12. Measuring the play of the side gears (Sec. 6)

Fig. 8.13. Installing the pinion gear shaft locking pin (Sec. 6)

Fig. 8.14. Fitting the collapsible spacer (Sec. 6)

Fig. 8.15. Fitting an adjustable cup (Sec. 6)

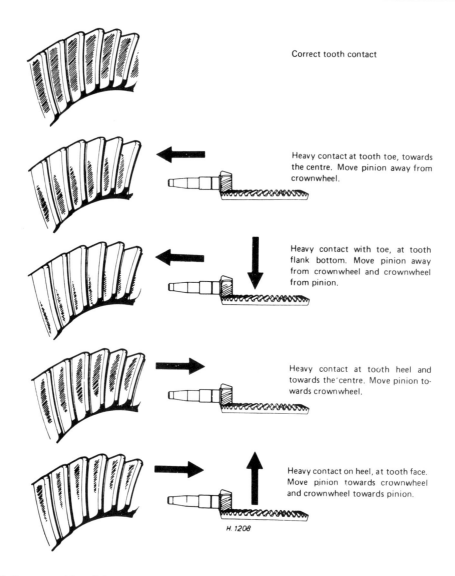

Fig. 8.16. Correct meshing of the crownwheel and pinion and repositioning guide for incorrect tooth meshing (Sec. 6)

setting.

45 The best check the D-I-Y motorist can make to ascertain the correct meshing of the crownwheel and pinion is to smear a little engineer's blue onto the crownwheel and then rotate the pinion. The contact mark should appear right in the middle of the crownwheel teeth. If the mark appears on the toe or the heel of the crownwheel teeth then the crownwheel must be moved closer or further away from the pinion. The various tooth patterns that may be obtained are illustrated (Fig. 8.16).

46 When the correct meshing between the crownwheel and pinion has been obtained refit the adjustment cup lock plates, bolts and spring washers.

47 The differential unit can now be refitted to the axle casing.

7 Pinion oil seal - removal and replacement

1 Jack up the rear of the car and secure on stands both under the body frame and axle casing.

2 Remove the roadwheels and brake drums.

3 Disconnect the propeller shaft from the pinion drive flange (refer to Chapter 7, if necessary.)

4 Using either a spring balance and a length of cord wrapped round the drive pinion or a torque wrench (lb f in) check and record the turning torque of the pinion.

5 Hold the drive pinion quite still with a suitable tool and remove the pinion self-locking nut.

6 Remove the washer, drive flange and dust deflector, and then prise out the oil seal. Do not damage or lever against the pinion shaft splines, during this operation.

7 Tap in the new oil seal using a piece of tubing as a drift. Do not inadvertently knock the end of the pinion shaft.

8 Repeat the operations described in paragraphs 37 to 40 of Section 6, but ensuring that the final pinion turning torque figure agrees with that recorded before dismantling.

9 Refit the brake drums, propeller shaft and roadwheels and lower the car.

10 Finally check and top up the oil level in the axle if necessary.

8 Fault diagnosis - rear axle

Symptom	Reason/s
Vibration	Worn axleshaft bearings. Loose drive flange bolts. Out of balance propeller shaft. Wheels require balancing.
Noise	Insufficient lubricant. Worn gears and differential components generally.
'Clunk' on acceleration or deceleration	Incorrect crownwheel and pinion mesh. Excessive backlash due to wear in crownwheel and pinion teeth. Worn axleshaft or differential side gear splines. Loose drive flange bolts. Worn drive pinion flange splines. Loose roadwheel nuts.
Oil leakage	Faulty pinion or axleshaft oil seals. May be caused by blocked axle housing breather.

Chapter 9 Braking system

For modifications, and information applicable to later models, see Supplement at end of manual

Contents

Bleeding the hydraulic system ...	12
Drum brake backplate - removal and refitting ...	8
Drum brake shoes - inspection and renewal ...	6
Drum brake wheel cylinder - removal, inspection, servicing and refitting ...	7
Fault diagnosis - braking system ...	22
Flexible hoses - inspection, removal and refitting ...	11
Front brake caliper - removal and refitting ...	3
Front brake caliper - servicing ...	4
Front disc pads - inspection and renewal ...	2
Front disc (rotor) and hub - removal and refitting ...	5
General description ...	1
Handbrake cable and rod (later models) - removal and refitting ...	17
Handbrake cables (early models) - removal and refitting ...	16
Handbrake control lever - removal and refitting ...	18
Handbrake (early models) - adjustment ...	14
Handbrake (later models) - adjustment ...	15
Master cylinder - removal and refitting ...	9
Master cylinder - servicing ...	10
Pressure differential valve - description and servicing ...	13
Vacuum servo unit - description ...	19
Vacuum servo unit - removal and refitting ...	20
Vacuum servo unit - servicing ...	21

Specifications

General
System type	Dual line, hydraulic
Front brakes	Disc, self-adjusting
Rear brakes	Drum, self-adjusting
Handbrake (parking brake)	Self-adjusting, cable operated to rear wheels

Front brakes
Disc diameter:	
Inner	5.5 in (139.7 mm)
Outer	9.5 in (241.3 mm)
Disc thickness:	
Nominal	0.376 in (9.54 mm)
Disc run out (max. including hub)	0.0019 in (0.05 mm)
Wheel cylinder diameter	1.89 in (48.1 mm)
Total swept brake area (two wheels)	182 in^2 (1175 cm^2)

Rear brakes
Drum diameter	7.99 to 8.00 in (202.95 to 203.2 mm)
Shoe width	1.45 in (36.8 mm)
Wheel cylinder diameter	0.75 in (19.05 mm)
Total swept area (two wheels)	72.7 in^2 (469 cm^2)
Rear brake linings - minimum permissible thickness (bonded linings)	1/32 in (0.03 in/0.8 mm)

Master cylinder diameter
Without servo	0.70 in (17.78 mm)
With servo	0.81 in (20.64 mm)

Servo boost ratio (where fitted)
4.3 : 1

Disc pads
Material	Ferodo 2434F/1D 334
Minimum permissible thickness	1/8 in (0.125 in/3.2 mm)

Torque wrench settings
	lb f ft	kg f m
Caliper to front suspension unit	4.8 to 6.9	35 to 50
Brake disc to hub	4.2 to 4.7	30 to 34
Brake backplate to axle housing	2.1 to 2.5	15 to 18
Hydraulic unions	0.7 to 1.0	5 to 7

Chapter 9/Braking system

1 General description

Disc brakes are fitted to the front wheels and single leading shoe drum brakes at the rear. The mechanically operated handbrake works on the rear wheels only.

The brakes fitted to the front wheels are of the rotating disc and static caliper type, with one caliper per disc, each caliper containing two piston operated friction pads, which on application of the footbrake pinch the disc rotating between them. The front brakes are of the trailing caliper type to minimise the entry of water.

Application of the footbrake creates hydraulic pressure in the master cylinder and fluid from the cylinder travels via steel and flexible pipes to the cylinders in each half of the calipers, thus pushing the pistons to which are attached the friction pads, into contact with either side of the disc.

Two seals are fitted to the operating cylinders, the outer seal prevents moisture and dirt entering the cylinder, while the inner seal which is retained in a groove inside the cylinder, prevents fluid leakage.

As the friction pads wear so the pistons move further out of the cylinders and the level of fluid in the hydraulic reservoir drops. Disc pad wear is therefore taken up automatically and eliminates the need for periodic adjustment by the owner.

A floor mounted handbrake (parking brake) lever is located between the front seats.

On early models, a single cable runs from the lever to a compensator mechanism on the back of the rear axle casing. From the compensator a single cable runs to the rear brake drums. As the rear brake shoes wear the handbrake cables operate a self adjusting mechanism in the rear brake drums thus doing away with the necessity for the owner to adjust the brakes on each rear wheel individually.

On later models, a cable runs from the handbrake lever through an abutment bracket on the rear axle to the right-hand brake backplate. A transverse rod connects from the abutment to the left-hand brake backplate.

The dual line braking system has a separate hydraulic circuit for the front and rear brakes, so that if failure of the hydraulic system to the front or rear brakes should occur, half the braking system will still operate.

In the event of a failure to the front or rear circuit, it is imperative that the fault be repaired at the earliest opportunity.

Some models are fitted with a differential valve which, when a failure occurs, actuates a warning light located in the facia panel. Also fitted to some models is a servo unit. This reduces the effort required by the driver to operate the brakes under all operating conditions.

2 Front disc pads - inspection and renewal

1 Apply the handbrake, remove the front wheel trim (where applicable), slacken the wheel nuts, jack up the front of the car and place on firmly based axle stands. Remove the front wheel.
2 Inspect the amount of friction material left on the pads. The pads must be renewed when the thickness of the friction material has been reduced to a minimum of 1/8 inch (3 mm). Should disc pad renewal be necessary, always replace as a pair and do not interchange the pads or discs.
3 If the fluid level in the master cylinder reservoir is high, when the pistons are moved into their respective bores to accommodate new pads the level could rise sufficiently for the fluid to overflow. Place absorbent cloth around the reservoir or syphon a little fluid out so preventing paintwork damage being caused by the hydraulic fluid.
4 Using a pair of long nosed pliers, extract the two small clips that hold the main retaining pins in place (photo).
5 Remove the main retaining pins which run through the caliper and the metal backing of the pads and the shims (Fig. 9.1).
6 The disc pads can now be removed from the caliper. If they prove difficult to remove by hand a pair of long nosed pliers can be used.
7 Carefully clean the recesses in the caliper in which the disc pads and shims lie, and the exposed faces of each piston, from all traces of dirt or rust.
8 Using a piece of wood carefully retract the pistons.
9 Place the tension springs on the disc pads and shims and locate in the caliper. Insert the main pad retaining pins making sure that the tangs of the tension springs are under the retaining pins. Secure the pins with the small wire clips (Fig. 9.2).

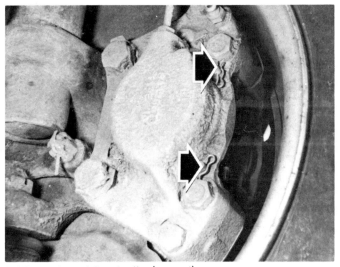

2.4 The main retaining pin clips (arrowed)

10 Refit the roadwheel and lower the car. Tighten the wheel nuts securely and replace the wheel trim.
11 To correctly seat the pistons pump the brake pedal several times and finally top up the hydraulic fluid level in the master cylinder reservoir as necessary.

3 Front brake caliper - removal and refitting

1 Apply the handbrake, remove the front wheel trim, slacken the wheel nuts, jack up the front of the car and place on firmly based axle stands. Remove the front wheel.
2 Wipe the top of the master cylinder reservoir and unscrew the cap. Place a piece of polythene sheet over the top of the reservoir and refit the cap.
3 Remove the disc pads, as described in Section 2.
4 If it is intended to fit new caliper pistons and/or seals, depress the brake pedal to bring the pistons into contact with the disc and assist subsequent removal of the pistons.
5 Wipe the area clean around the flexible hose bracket and detach the pipes as described in Section 11. Tape up the end of the pipe to stop the possibility of dirt ingress.
6 Using a screwdriver or chisel bend back the tabs on the locking plate and undo the two caliper body mounting bolts. Lift away the caliper from its mounting flange on the suspension leg (Fig. 9.3).
7 To refit the caliper, position it over the disc and move it until the mounting bolt holes are in line with the two front holes in the suspension leg mounting flange.
8 Fit the caliper retaining bolts through the two holes in a new locking plate and insert the bolts through the caliper body. Tighten the bolts to the specified torque wrench setting.
9 Using a screwdriver, pliers or chisel bend up the locking plate tabs so as to lock the bolts.
10 Remove the tape from the end of the flexible hydraulic pipe and reconnect it to the union on the hose bracket. Be careful not to cross the thread of the union nut during the initial turns. The union nut should be tightened securely using a spanner of short length.
11 Push the pistons into their respective bores so as to accommodate the pads. Watch the level of hydraulic fluid in the master cylinder reservoir as it can overflow if too high whilst the pistons are being retracted. Place absorbent cloth around the reservoir or syphon a little fluid out so preventing paintwork damage.
12 Fit the pads, shims and tension springs, as described in Section 2.
13 Bleed the hydraulic system, as described in Section 12. Replace the roadwheel and lower the car.

4 Front brake caliper - servicing

1 The pistons should be removed first. To do this, half withdraw one

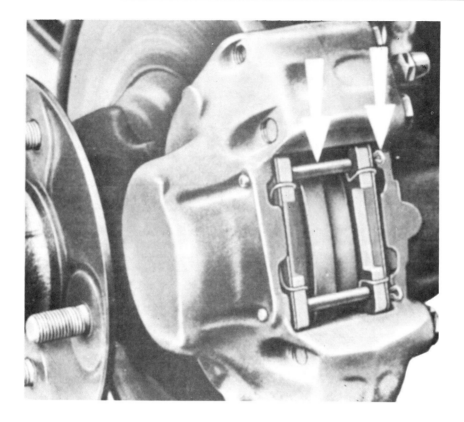

Fig. 9.1. Brake pad retaining pins and clips (arrowed) (Sec. 2)

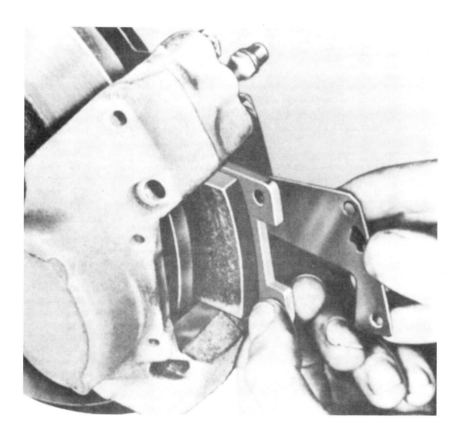

Fig. 9.2. Fitting brake pads and shims (Sec. 2)

Chapter 9/Braking system 97

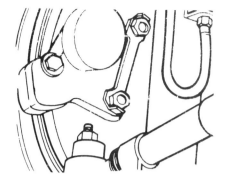

Fig. 9.3. Brake caliper mounting bolts (Sec. 3)

piston from its bore in the caliper body.
2 Carefully remove the securing circlip and extract the sealing boot from its location in the lower part of the piston skirt. Completely remove the piston.
3 If difficulty is experienced in withdrawing the piston use a jet of compressed air or a foot pump to move it out of its bore.
4 Remove the sealing bellows from its location in the annular ring which is machined in the cylinder bore.
5 Remove the piston sealing ring from the cylinder bore using a small screwdriver but do take care not to scratch the fine finish of the bore.
6 To remove the second piston repeat paragraphs 1 to 5 inclusive.
7 It is important that the two halves of the caliper are not separated under any circumstances. If hydraulic fluid leaks are evident from the joint, the caliper must be renewed complete.
8 Thoroughly wash all parts in methylated spirits or clean hydraulic fluid. During reassembly new rubber seals must be fitted, these should be well lubricated with clean hydraulic fluid.
9 Inspect the pistons and bores for signs of wear, score marks or damage and, if evident, new parts should be obtained ready for fitting or a new caliper obtained.
10 To reassemble, fit one of the piston seals into the annular groove in the cylinder bore.
11 Fit the rubber boot to the cylinder bore groove so that the lip is turned outward.
12 Lubricate the seal and rubber boot with clean hydraulic fluid. Push the piston, crown first, through the rubber sealing bellows and then into the cylinder bore. Take care as it is easy for the piston to damage the rubber boot.
13 With the piston half inserted into the cylinder bore fit the inner edge of the boot into the annular groove in the piston skirt.
14 Push the piston down the bore as far as it will go. Secure the rubber boot to the caliper with the circlip.
15 Repeat paragraphs 10 to 14 inclusive for the second piston.
16 The caliper is now ready for refitting. It is recommended that the hydraulic pipe end is temporarily plugged to stop any dirt entering whilst it is being refitted, before the pipe connection is made.

5 Front disc (rotor) and hub - removal and refitting

Note: Brake discs (rotors) are fitted as matched pairs and should therefore never be renewed or reground as single items.
1 After jacking up the car and removing the front wheel, remove the caliper, as described in Section 3.
2 Tap off the dust cap from the centre of the hub.
3 Remove the split pin from the nut retainer and lift the retainer away.
4 Unscrew the adjusting nut and lift away the thrust washer and outer tapered bearing.
5 Pull off the complete hub and disc assembly from the stub axle.
6 From the back of the hub assembly carefully prise out the grease seal and lift away the inner tapered bearing.
7 Carefully clean out the hub and wash the bearings with petrol making sure that no grease or oil is allowed to get onto the brake disc.
8 Should it be necessary to separate the disc from the hub for renewal or regrinding, first bend back the locking tabs and undo the four securing bolts. With a scriber mark the relative positions of the hub and disc to ensure refitting in their original positions and separate the disc from the hub.
9 Thoroughly clean the disc and inspect for signs of deep scoring, cracks or excessive corrosion. If these are evident, the discs may be reground but no more than a maximum total of 0.060 inch (1.524 mm) may be removed. It is however, desirable to fit new discs if at all possible.
10 To reassemble make quite sure that the mating faces of the disc and hub are very clean and place the disc on the hub, lining up any previously made marks.
11 Fit the four securing bolts and two new tab washers and tighten the bolts in a progressive and diagonal manner to the specified torque wrench setting. Bend up the locking tabs.
12 Work some grease well into the bearing, fully pack the bearing cages and rollers. Note: Leave the hub and grease seal empty to allow

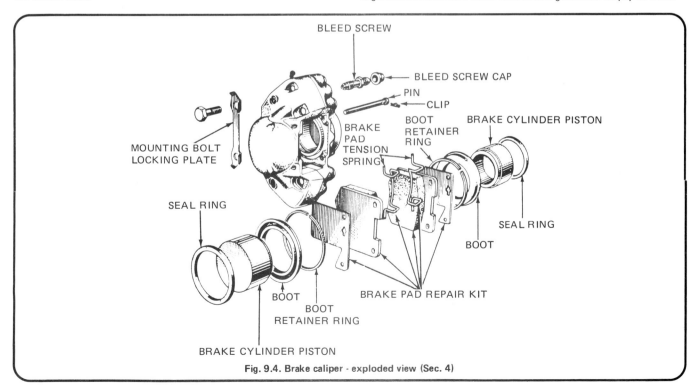

Fig. 9.4. Brake caliper - exploded view (Sec. 4)

for subsequent expansion of the grease.

13 To reassemble the hub, first fit the inner bearing and then gently tap the grease seal into the hub. A new seal must always be fitted as, during removal, it was probably damaged or distorted. The lip must face inwards to the hub.

14 Replace the hub and disc assembly onto the stub axle and slide in the outer bearing and thrust washer.

15 Refit the adjusting nut and tighten it to a torque wrench setting of 27 lb f ft (3.7 kg f m) whilst rotating the hub and disc to ensure free movement and centralisation of the bearings. Slacken the nut back by 90° which will give the required endfloat of 0.001 - 0.005 in (0.03 - 0.13 mm). Fit the nut retainer and a new split pin, but at this stage do not lock the split pin.

16 If a dial indicator gauge is available, it is advisable to check the disc for run-out. The measurement should be taken as near to the edge of the worn yet smooth part of the disc as possible, and must not exceed 0.002 in (0.05 mm). If the figure obtained is found to be excessive, check the mating surfaces of the disc and hub for dirt or damage and check the bearing and cups for excessive wear or damage.

17 If a dial indicator gauge is not available the run-out can be checked by means of a feeler gauge placed between the casting of the caliper and the disc. Establish a reasonably tight fit with the feeler gauge between the top of the casting and the disc and rotate the disc and hub. Any high or low spots will immediately become obvious by extra tightness or looseness of the fit of the feeler gauge. The amount of run-out can be checked by adding or subtracting feeler gauges as necessary.

18 Once the disc run-out has been checked and found to be correct, bend the ends of the split pin back and replace the dust cap.

19 Reconnect the brake hydraulic pipe and bleed the brakes as described in Section 12 of this Chapter.

6 Drum brake shoes - inspection and renewal

After high mileages, it will be necessary to fit replacement shoes with new linings. Refitting new brake linings to shoes is not considered economic, or possible, without the use of special equipment. However, if the services of a local garage or workshop having brake relining equipment are available then there is no reason why the original shoes should not be relined successfully. Ensure that the correct specification linings are fitted to the shoes.

1 Chock the front wheels, jack up the rear of the car and place on firmly based axle stands. Remove the roadwheel.

2 Release the brake drum retaining screw, and using a soft-faced mallet on the outer circumference of the brake drum remove the brake drum (photo).

3 The brake linings should be renewed if they are so worn that the rivet heads are flush with the surface of the lining. If bonded linings are fitted, they must be renewed when the lining material has worn down to the minimum specified thickness.

4 Depress each shoe holding down spring and rotate the spring retaining washer through 90° to disengage it from the pin secured to the backplate. Lift away the washer and spring.

5 Ease each shoe from its location slot in the fixed pivot and then detach the other end of each shoe from the wheel cylinder.

6 Note which way round and into which holes in the shoes the two retracting springs fit and detach the retracting springs.

7 Lift away the two brake shoes and retracting springs.

8 If the shoes are to be left off for a while, place a warning on the steering wheel. Also place an elastic band round the wheel cylinder to stop the piston falling out.

9 Withdraw the ratchet wheel assembly from the wheel cylinder and rotate the wheel until it abuts the slot head bolt shoulder. If this is not done difficulty will arise in refitting the brake drum.

10 Thoroughly clean all traces of dust from the shoes, backplates and brake drums using a stiff brush. Brake dust can cause judder, or squeal and, therefore, it is important to clean out as described. It is not recommended that compressed air is used as it blows up dust which should not be inhaled.

11 Check that the piston is free in the cylinder, that the rubber dust covers are undamaged and in position, and that there are no hydraulic leaks.

12 Prior to reassembly smear a trace of brake grease on the shoe support pads, brake shoe pivots and on the ratchet wheel face and threads.

13 To reassemble first fit the retracting springs to the shoe webs in the same position as was noted during removal.

14 Fit the shoe assembly to the backplate by first positioning the rear shoe in its location on the fixed pivot and over the handbrake link. Follow this with the front shoe.

15 Secure each shoe to the backplate with the spring and dished washer, dish facing inwards, and turning through 90° to lock in position. Make sure that each shoe is firmly seated on the backplate.

16 Refit the brake drum and push it up the studs as far as it will go. Secure with the retaining screw.

17 The shoes must next be centralised by the brake pedal being depressed firmly several times.

18 Pull on and then release the handbrake several times to reset the adjuster mechanism. It is important to note that with the ratchet wheel in the fully off adjustment position, it is possible for the indexing lever on the handbrake link to over-ride the ratchet and stay in this position. When operating the link lever it is necessary to ensure that it always returns to the fully off position each time.

19 Refit the roadwheel and lower the car. Road test to ensure correct operation of the brakes.

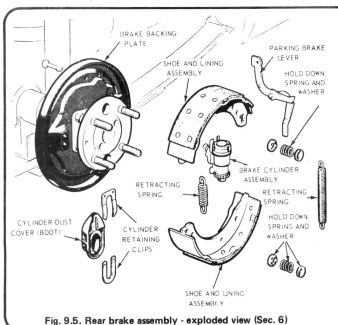

Fig. 9.5. Rear brake assembly - exploded view (Sec. 6)

6.2 Rear brake with drum removed

7 Drum brake wheel cylinder - removal, inspection, servicing and refitting

1 Refer to Section 6 and remove the brake drum and shoes. Clean down the rear of the backplate using a stiff brush. Place a quantity of rag under the backplate to catch any hydraulic fluid that may issue from the open pipe or wheel cylinder.
2 Wipe the top of the brake master cylinder reservoir and unscrew the cap. Place a piece of polythene sheet over the top of the reservoir and replace the cap.
3 Using an open ended spanner carefully unscrew the hydraulic pipe connection union at the rear of the wheel cylinder. To prevent dirt entering, tape over the end of the pipe.
4 Withdraw the split pin and clevis pin from the handbrake lever at the rear of the backplate.
5 Using a screwdriver carefully ease the rubber dust cover from the rear of the backplate and lift away.
6 Pull off the two 'U' shaped retainers holding the wheel cylinder to the backplate noting that the spring retainer is fitted from the handbrake link end of the wheel cylinder and the flat retainer from the other end, the flat retainer being located between the spring retainer and the wheel cylinder.
7 The wheel cylinder and handbrake link can now be removed from the brake backplate.
8 To dismantle the wheel cylinder first remove the small metal clip holding the rubber dust cap in place then prise off the dust cap.
9 Take the piston complete with its seal out of the cylinder bore and then withdraw the spring. Should the piston and seal prove difficult to remove gentle pressure will push it out of the bore.
10 Inspect the cylinder bore for score marks caused by impurities in hydraulic fluid. If any are found the cylinder and piston will require renewal together, as a replacement unit.
11 If the cylinder bore is sound thoroughly clean it out with fresh hydraulic fluid.
12 The old rubber seal will probably be visibly worn or swollen. Detach it from the piston, smear a new rubber seal with hydraulic fluid and assemble it to the piston with the flat face of the seal next to the piston rear shoulder.
13 Reassembly is a direct reversal of the dismantling procedure. If the rubber dust cap appears to be worn or damaged it should also be renewed.
14 Before commencing refitting smear the area where the cylinder slides on the backplate and the brake shoe support pads, brake shoe pivots, ratchet wheel face and threads with brake grease.
15 Refitting is a straightforward reversal of the removal sequence but the following parts should be checked with extra care.
16 After fitting the rubber boot, check that the wheel cylinder can slide freely in the backplate and that the handbrake link operates the self-adjusting mechanism correctly.
17 It is important to note that the self-adjusting ratchet mechanism on the right-hand rear brake is right-hand threaded and the mechanism on the left-hand rear brake is left-hand threaded.
18 When refitting is complete, bleed the braking system as described in Section 12.

8 Drum brake backplate - removal and refitting

1 To remove the backplate refer to Chapter 8 and remove the halfshaft.
2 Detach the handbrake cable from the handbrake relay lever on the backplate.
3 Wipe off the top of the master cylinder reservoir and unscrew the cap. Place a piece of polythene sheet over the top of the reservoir and replace the cap.
4 Using an open ended spanner, carefully unscrew the hydraulic pipe connection union to the rear of the wheel cylinder. To prevent dirt entering tape over the pipe ends.
5 The brake backplate may now be lifted away.
6 Refitting is the reverse sequence to removal. It will be necessary to bleed the brake hydraulic system, as described in Section 12.

9 Master cylinder - removal and refitting

1 Apply the handbrake and chock the front wheels. Drain the fluid from the master cylinder reservoir and master cylinder by attaching a plastic bleed tube to one of the front brake bleed screws. Undo the screw one turn and then pump the fluid out into a clean glass container by means of the brake pedal. Hold the brake pedal against the floor at the end of each stroke and tighten the bleed screw. When the pedal has returned to its normal position loosen the bleed screw and repeat the process. The above sequence should now be carried out on one of the rear brake bleed screws.
2 Wipe the area around the two union nuts on the side of the master cylinder body and using an open ended spanner undo the two union nuts. Tape over the ends of the pipes to stop dirt entering.
3 From inside the car remove the split pin and washer and push out the pin securing the master cylinder pushrod to the brake pedal. Undo and remove the two nuts and spring washers that secure the master cylinder to the bulkhead. Lift away the master cylinder and ensure that no hydraulic fluid is allowed to drip onto the paintwork. On models fitted with a servo it is only necessary to undo and remove the two nuts securing the master cylinder to the servo, after undoing the unions.
4 Refitting the master cylinder is the reverse sequence to removal. Always start the union nuts before finally tightening the master cylinder nuts. It will be necessary to bleed the complete hydraulic system: full details will be found in Section 12.
5 If a replacement master cylinder is to be fitted, it will be necessary to lubricate the seals before fitting to the car as they have a protective coating when originally assembled. Remove the blanking plugs from the hydraulic pipe union seatings. Inject clean hydraulic fluid into the

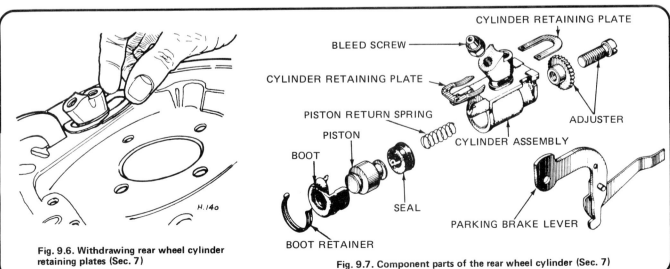

Fig. 9.6. Withdrawing rear wheel cylinder retaining plates (Sec. 7)

Fig. 9.7. Component parts of the rear wheel cylinder (Sec. 7)

10 Master cylinder - servicing

If the master cylinder is to be dismantled after removal, proceed as follows:
1 The component parts are shown in Fig. 9.8.
2 Prior to dismantling wipe the exterior of the master cylinder clean.
3 Undo and remove the two screws and spring washers holding the reservoir to the master cylinder body. Lift away the reservoir. Using an Allen key, or wrench, unscrew the tipping valve nut and lift away the seal. Using a suitable diameter rod push the primary plunger down the bore, this operation will enable the tipping valve to be withdrawn.
4 Using a compressed air jet, very carefully applied to the rear outlet connection, blow out all the master cylinder internal components. Alternatively, shake out the parts. Take care that adequate precautions are taken to ensure all parts are caught as they emerge.
5 Separate the primary and secondary plungers from the intermediate spring. Use the fingers to remove the gland seal from the primary plunger.
6 The secondary plunger assembly should be separated by lifting the thimble leaf over the shouldered end of the plunger. Using the fingers, remove the seal from the secondary plunger.
7 Depress the secondary spring, allowing the valve stem to slide through the keyhole in the thimble, thus releasing the tension on the spring.
8 Detach the valve spacer, taking care of the spring washer which will be found located under the valve head.
9 Thoroughly wash all parts in either methylated spirits or clean approved hydraulic fluid and place in order ready for inspection.
10 Examine the bores of the master cylinder carefully for any signs of scoring, ridges or corrosion. If there is any doubt as to the condition of the bore, then a new assembly must be obtained.
11 All components should be assembled wet by dipping in clean brake fluid. Using fingers only, fit new seals to the primary and secondary plungers ensuring that they are the correct way round. Place the dished washer with the dome against the underside of the valve seat. Hold it in position with the valve spacer ensuring that the legs face towards the valve seal.
12 Replace the plunger return spring centrally on the spacer, insert the thimble into the spring, and depress until the valve stem engages in the keyhole of the thimble.
13 Insert the reduced end of the plunger into the thimble until the thimble engages under the shoulder of the plunger, and press home the thimble leaf. Replace the intermediate spring between the primary and secondary plungers.
14 Check that the master cylinder bore is clean and smear with clean brake fluid. With the complete assembly suitably lubricated, carefully insert the assembly into the bore. Ease the lips of the piston seals into the bore taking care that they do not roll over. Push the assembly fully home.
16 Refit the tipping valve assembly and seal to the cylinder bore, and tighten the securing nut to a torque wrench setting of 27 to 35 lb f ft (4.8 to 6.22 kg f m).
17 Using a clean screwdriver push the primary piston in and out checking that the recuperating valve opens when the screwdriver is withdrawn and closes again when it is pushed in.
18 Check the condition of the front and rear reservoir gaskets and if there is any doubt as to their condition they must be renewed.
19 Replace the hydraulic fluid reservoir and tighten the two retaining screws.
20 The master cylinder is now ready for refitting. Bleed the complete hydraulic system and road test the car.

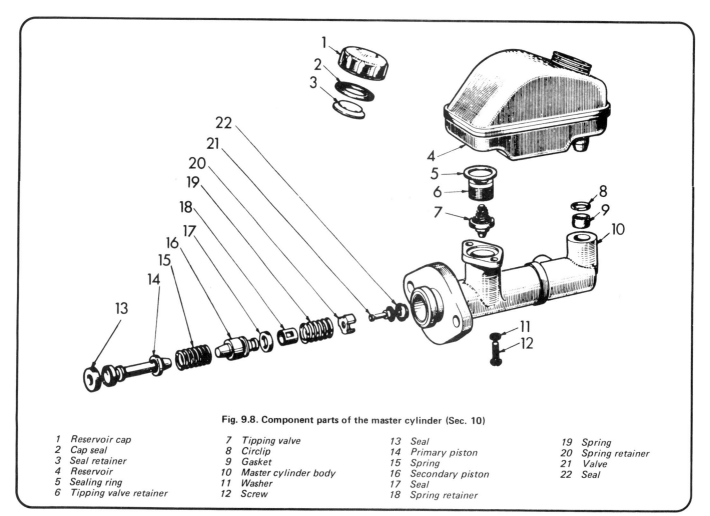

Fig. 9.8. Component parts of the master cylinder (Sec. 10)

1 Reservoir cap
2 Cap seal
3 Seal retainer
4 Reservoir
5 Sealing ring
6 Tipping valve retainer
7 Tipping valve
8 Circlip
9 Gasket
10 Master cylinder body
11 Washer
12 Screw
13 Seal
14 Primary piston
15 Spring
16 Secondary piston
17 Seal
18 Spring retainer
19 Spring
20 Spring retainer
21 Valve
22 Seal

11 Flexible hose - inspection, removal and refitting

1 Inspect the condition of the flexible hydraulic hoses leading from under the front wings to the brackets on the front suspension units, and also the single hose on the rear axle casing. If they are swollen, damaged or chafed, they must be renewed.
2 Undo the locknuts at both ends of the flexible hoses and then holding the hexagon nut on the flexible hose steady undo the other union nut and remove the flexible hose and washer.
3 Refitting is a reversal of the removal procedure, but carefully check that all the securing brackets are in a sound condition and that the locknuts are tight.

12 Bleeding the hydraulic system

1 Removal of all the air from the hydraulic system is essential to the correct working of the braking system, and before undertaking this, examine the fluid reservoir cap to ensure that both vent holes, one on the exterior, on the top, and the second underneath inside the cap, but not in line, are clear, check the level of fluid and top-up if required.
2 Check all brake line unions and connections for possible seepage, and at the same time check the condition of the rubber hoses, which may be perished.
3 If the condition of the wheel cylinders is in doubt, check for possible signs of fluid leakage.
4 If there is any possibility of incorrect fluid having been put into the system, drain all the fluid out and flush through with methylated spirits. Renew all piston seals and cups since these will be affected and could possibly fail under pressure.
5 Gather together a clean jar, a 9 inch (230 mm) length of tubing which fits tightly over the bleed nipples, and a tin of the correct brake fluid.
6 Clean the dirt from around the front caliper bleed nipple which is furthest from the master cylinder. Fit one end of the tubing over the bleed nipple and put the other into the jar containing a little brake fluid. Ensure that this end remains immersed during the bleeding process.
7 Open the bleed valve with a spanner and then have an assistant quickly depress the brake pedal. After slowly releasing the pedal, hold for a moment to allow the fluid to recoup in the master cylinder and then depress again. This will force air from the system. Continue until no air bubbles can be seen coming from the tube. At intervals make certain that the reservoir is kept topped up, otherwise air will enter at this point again.
8 Repeat this operation on the other front brake and the rear brakes (one bleed nipple only for the rear brakes). When completed, check the level of the fluid in the reservoir and then check the feel of the brake pedal, which should be firm and free from any 'spongy' action, which is normally associated with air in the system.

13 Pressure differential valve - description and servicing

This device is incorporated in the hydraulic circuit on some models. It is a valve in which a piston is kept 'in balance' when the hydraulic pressure in the independent front and rear circuits are equal. In the event of a drop in pressure in either circuit, the piston is displaced and makes an electrical contact to illuminate a warning light on the instrument panel.
1 To dismantle the valve, first disconnect the hydraulic pipes at their unions on the body. To prevent a loss of hydraulic fluid either place a piece of polythene under the cap of the master cylinder and screw it down tightly or plug the ends of the two pipes leading from the master cylinder.
2 Referring to Fig. 9.9, disconnect the wiring from the warning switch.
3 Undo the single bolt holding the assembly to the rear of the engine compartment and remove it from the car.
4 To dismantle the assembly start by undoing the end plug and discarding the gasket.
5 Unscrew the switch assembly from the top of the unit then push the piston out of the bore taking extreme care not to damage the bore during this operation.
6 Take the small seals from the piston followed by the sleeves.
7 Carefully examine the piston and bore for score marks, scratches or damage; if any are found the complete unit must be exchanged for a new one. Also check that the piston retaining clips are secure and undamaged.
8 Reassembly of the unit is the reverse of the removal procedure, ensuring that all parts are adequately lubricated with hydraulic brake fluid.

14 Handbrake (early models) - adjustment

1 Adjustment of the handbrake is normally automatically carried out by the action of the rear brake adjusters and the only periodical attention required is to lubricate the cable pivot points (Fig. 9.10). When new components have been fitted or where the handbrake cable has stretched, then the following operations should be carried out.
2 Chock the front wheels, jack up the rear of the car and support on firmly based axle stands. Release the handbrake.
3 Slide under the car and check that the primary cable follows its correct run and is correctly in its guide. The cable guides must be kept well greased at all times.
4 First adjust the effective length of the primary cable by slackening the locknut on the end of the cable adjacent to the relay lever on the rear axle (Fig. 9.11).
5 Adjust the nut until the primary cable has no slack in it and the relay lever is just clear of the slot in the banjo casing. Retighten the locknut.

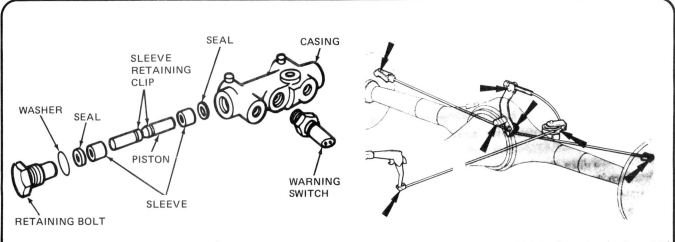

Fig. 9.9. Pressure differential valve - exploded view (Sec. 13)

Fig. 9.10. Handbrake cable layout and lubrication points (early models) (Sec. 14)

Chapter 9/Braking system

Fig. 9.11. Primary cable adjustment point (early models) (Sec. 14)

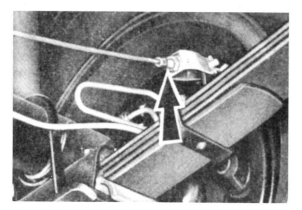

Fig. 9.12. Transverse cable adjustment point (early models) (Sec. 14)

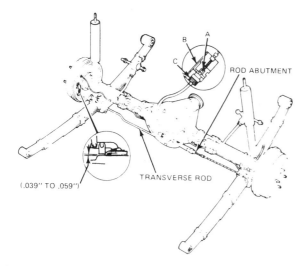

Fig. 9.13. Handbrake assembly layout and adjustment points (later models) (Sec. 15)

6 Slacken the locknut on the end of the transverse cable adjacent to the right-hand rear brake (Fig. 9.12). Check that the parking brake operating levers are in the fully 'off' position, that is back on their stops, and adjust the cable so that there is no slack. Check that the operating levers are still on their stops and tighten the locknut.
7 Lower the car to the ground.

15 Handbrake (later models) - adjustment

1 Adjustment of the handbrake is normally automatically carried out by the action of the rear brake adjusters. When new components have been fitted or where the handbrake cable has stretched, then the

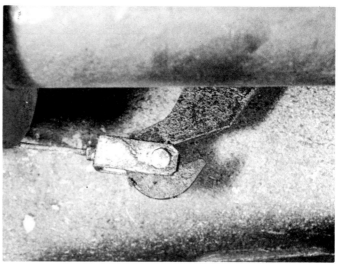

16.3 The hooked end of the brake primary cable

following operations should be carried out.
2 Chock the front wheels, jack up the rear of the car and support on firmly based axle stands. Release the handbrake.
3 First ensure that the primary cable is properly located, then engage the keyed sleeve 'A' into the abutment slot 'B' (Fig. 9.13).
4 Turn the adjuster nut 'C' until all cable slack is eliminated, and a clearance of 0.039 to 0.059 in (1 to 1.5 mm) exists between the parking brake lever stop and the brake backplate.
5 Lower the car to the ground.

16 Handbrake cables (early models) - removal and refitting

Primary cable

1 Chock the front wheels, jack up the rear of the car and support on firmly based axle stands. Release the handbrake.
2 Working under the car unscrew and remove the nuts that secure the end of the primary cable to the relay lever located at the rear of the axle casing.
3 Detach the primary cable from the end of the handbrake lever by removing the split pin and withdrawing the clevis pin. Note: On some models the cable hooks onto the end of the handbrake lever (photo).
4 Detach the cable from its underbody guides and lift away.
5 Refitting the primary cable is the reverse sequence to removal but the following additional points should be noted:

   a) Apply some grease to the cable guides and insert the cable. Also lubricate the front clevis pin (if fitted).
   b) Refer to Section 14 and adjust the primary cable.

Transverse cable

1 Chock the front wheels, jack up the front of the car and support on firmly based axle stands. Release the handbrake.
2 Working under the car remove the split pin and withdraw the clevis pin that secures the transverse cable to the left-hand backplate.
3 Detach the cable from the right-hand rear backplate by removing the locknut and unscrewing the cable from the clevis.
4 Remove the split pin from the pulley pin, and withdraw the pulley pin. Lift away the little pulley wheel and transverse cable.
5 Refitting the transverse cable is the reverse sequence to removal but the following additional points should be noted:

   a) Apply some grease to the pulley and pivot pin, the threaded end of the cable and the clevis pin.
   b) Adjust the transverse cable as described in Section 14.

17 Handbrake cable and rod (later models) - removal and refitting

Primary cable

1 Chock the front wheels, jack up the rear of the car and support on

firmly based axle stands. Release the handbrake.
2 Remove the spring clip and clevis pin connecting the handbrake cable to the handbrake lever.
3 Remove the spring clip and clevis pin from the right-hand rear brake lever, disconnect the cable.
4 Remove the handbrake cable-to-transverse rod retaining clip, then slide the cable clear of the rod bracket.
5 Slide the cable, adjusting nut and guide clear of the abutment bracket, and remove the assembly from the car.
6 Refitting is the reverse of the removal procedure. Apply a little general purpose grease to the rubbing and pivoting parts, then finally check the adjustment (Section 15).

Transverse rod

7 Initially proceed as described in paragraphs 1 and 2.
8 Remove the spring retaining clip which secures the handbrake cable to the transverse rod, and slide the cable assembly clear.
9 Remove the spring clip and clevis pin, then disconnect the rod from the left-hand rear brake lever.
10 Slide the rod out of the bushing on the axle casing.
11 Refitting is the reverse of the removal procedure. Apply a little general purpose grease to the rubbing and pivoting parts, then finally check the adjustment (Section 15).

18 Handbrake control lever - removal and refitting

1 Chock the front wheels, jack up the rear of the car and support on firmly based axle stands. Release the handbrake.
2 Working inside the car remove the carpeting from around the area of the handbrake lever.
3 Models fitted with a console: Refer to Chapter 12 and remove the console.
4 Remove the split pin and withdraw the clevis pin that connects the primary cable to the lower end of the handbrake lever; this protrudes under the floor panels. Note: On some models the cable hooks onto the end of the handbrake lever (photo 16.3).
5 Undo and remove the six self-tapping screws which secure the handbrake lever rubber boot to the floor. Draw the rubber boot up the lever.
6 Undo and remove the two bolts that secure the handbrake lever assembly to the floor. Lift away the lever assembly.
7 Refitting the lever assembly is the reverse sequence to removal. The following additional points should be noted:

   a) *Apply some grease to the primary cable clevis pin (if fitted).*
   b) *Adjust the primary cable as described in Sections 14 or 15.*

19 Vacuum servo unit - description

1 A vacuum servo unit is fitted into the brake hydraulic circuit in series with the master cylinder, to provide assistance to the driver when the brake pedal is depressed. This reduces the effort required by the driver to operate the brakes under all braking conditions.
2 The unit operates by vacuum obtained from the induction manifold and comprises basically a booster diaphragm and non-return valve. The servo unit and hydraulic master cylinder are connected together so that the servo unit piston rod acts as the master cylinder pushrod. The driver's braking effort is transmitted through another pushrod to the servo unit piston and its built-in control system. The servo unit piston does not fit tightly into the cylinder but has a strong diaphragm to keep its edges in constant contact with the cylinder wall, so assuring an airtight seal between the two parts. The forward chamber is held under vacuum conditions created in the inlet manifold of the engine and, during periods when the brake pedal is not in use, the controls open a passage to the rear chamber so placing it under vacuum conditions as well. When the brake pedal is depressed, the vacuum passage to the rear chamber is cut off and the chamber exposed to atmospheric pressure. The consequent rush of air pushes the servo piston forward in the vacuum chamber and operates the main pushrod to the master cylinder.
3 The controls are designed so that assistance is given under all conditions and, when the brakes are not required, vacuum in the rear chamber is established when the brake pedal is released. All air from the atmosphere entering the rear chamber is passed through a small air filter.
4 Under normal operating conditions the vacuum servo unit is very reliable and does not require overhaul except at very high mileages. In this case it is far better to obtain a service exchange unit, rather than repair the original unit.

20 Vacuum servo unit - removal and refitting

1 Slacken the clip securing the vacuum hose to the servo unit; carefully draw the hose from its union.
2 Refer to Section 9 and remove the master cylinder.
3 Using a pair of pliers remove the spring clip in the end of the brake pedal to pushrod clevis pin. Lift away the clevis pin and the bushes.
4 Undo and remove the nuts and spring washers securing the servo unit mounting bracket to the bulkhead. Lift away the servo unit and bracket.
5 Undo and remove the four nuts and spring washers that secure the bracket to the servo unit.
6 Refitting the servo unit is the reverse sequence to removal. It will be necessary to bleed the brake hydraulic system as described in Section 12.

21 Vacuum servo unit - servicing

Thoroughly clean the outside of the unit using a stiff brush and wipe with a non-fluffy rag. It cannot be too strongly emphasised that cleanliness is important when working on the servo. Before any attempt is made to dismantle, refer to Fig. 9.14 where it will be seen that two items of equipment are required. Firstly, a base plate must be made to enable the unit to be safely held in a vice. Secondly, a lever must be made similar to the form shown. Without these items it is impossible to dismantle satisfactorily.

To dismantle the unit proceed as follows:
1 Refer to Fig. 9.14 and, using a file or scriber make a line across the two halves of the unit to act as a datum for alignment.
2 Fit the previously made base plate into a firm vice and attach the unit to the plate using the master cylinder studs.
3 Use a piece of long rubber hose and connect one end to the connector on the engine inlet manifold and the other end to the non-return valve. Starting the engine will create a vacuum in the unit so drawing the two halves together.
5 Rotate the lever in an anticlockwise direction until the front shell indentations are in line with the recesses in the rim of the rear shell. Then press the lever assembly down firmly whilst an assistant stops the engine and quickly removes the vacuum pipe from the inlet manifold connector. Depress the operating rod so as to release the vacuum, whereupon the front and rear halves should part. If necessary, use a soft faced hammer and lightly tap the front half to break the bond.
6 Lift away the rear shell followed by the diaphragm return spring,

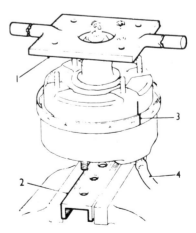

Fig. 9.14. Special tools required to dismantle servo unit (Sec. 21)

1 Lever
2 Base plate
3 Scribe line
4 Vacuum applied

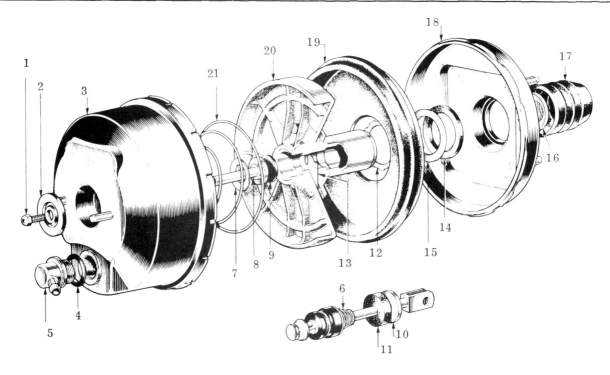

Fig. 9.15. Servo unit - component parts (Sec. 21)

1 Bolt	7 Dished washer	13 Stop key
2 Seat assembly	8 Brake servo pushrod	14 Seal
3 Front shell	9 Reaction disc	15 Piston guide
4 Seal	10 Washer	16 Filter retainer
5 Valve assembly	11 Filter	17 Dust cover
6 Pushrod assembly	12 Castellated washer	18 Rear shell

19 Diaphragm
20 Diaphragm plate
21 Spring

the dust cap, end cap and the filter. Also withdraw the diaphragm. Press down the valve rod and shake out the valve retaining plate. Then separate the valve rod assembly from the diaphragm plate (Fig. 9.15).

7 Gently ease the spring washer from the diaphragm plate and withdraw the pushrod and reaction disc.

8 The seal and plate assembly in the end of the front shell are a press fit. It is recommended that, unless the seal is to be renewed, they be left in-situ.

9 Thoroughly clean all parts. Inspect them for signs of damage, stripped threads etc., and obtain new ones as necessary. All seals should be renewed and for this a 'Major Repair Kit' should be purchased. This kit will also contain two separate greases which must be used as directed and not interchanged.

10 To reassemble, first smear the seal and bearing with Ford grease numbered '64949008 EM - 1C - 14' and refit the rear shell positioning it such that the flat face of the seal is towards the bearing. Press into position and refit the retainer.

11 Lightly smear the disc and hydraulic pushrod with Ford grease number '64949008 EM - 1C - 14'. Refit the reaction disc and pushrod to the diaphragm plate and press in the large spring washer. The small spring washer supplied in the 'Major Repair Kit' is not required. It is important that the length of the pushrod is not altered in any way and any attempt to move the adjustment bolt will strip the threads. If a new hydraulic pushrod has been required, the length will have to be reset. Details of this operation are given at the end of this Section.

12 Lightly smear the outer diameter of the diaphragm plate neck and the bearing surfaces of the valve plunger with Ford grease number '64949008 EM - 1C - 14'. Carefully fit the valve rod assembly into the neck of the diaphragm and fix with the retaining plate.

13 Fit the diaphragm into position and the non-return valve to the front shell. Next smear the seal and plate assembly with Ford grease numbered '6494008 EM - 1C - 15' and press into the front shell with the plate facing inwards.

14 Fit the front shell to the baseplate and the lever to the rear shell. Reconnect the vacuum hose to the non-return valve and the connector on the engine inlet manifold. Position the diaphragm return spring in the front shell. Lightly smear the outer bead of the diaphragm with Ford grease numbered '64949008 EM - 1C - 14' and locate the

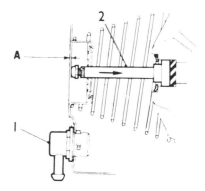

Fig. 9.16. Pushrod setting (Sec. 21)

A Pushrod setting 0.011 - 0.016 in (0.28 - 0.40 mm)
1 Vacuum applied
2 Pushrod against reaction disc

diaphragm assembly in the rear shell. Position the rear shell assembly on the return spring and line up the previously made scribe marks.

15 The assistant should start the engine. Watching one's fingers very carefully, press the two halves of the unit together and, using the lever tool, turn clockwise to lock the two halves together. Stop the engine and disconnect the hose.

16 Press a new filter into the neck of the diaphragm plate, refit the end cap and position the dust cover onto the special lugs of the rear shell.

17 Hydraulic pushrod adjustment only applies if a new pushrod has been fitted. It will be seen from Fig. 9.16 that there is a bolt screwed into the end of the pushrod. The amount of protrusion has to be adjusted in the following manner: Remove the bolt and coat the threaded portion with Loctite Grade B. Reconnect the vacuum hose to the adaptor on the inlet valve and non-return valve. Start the engine and screw the prepared bolt into the end of the pushrod. Adjust the

Chapter 9/Braking system

position of the bolt head so that it is 0.011 to 0.016 inch (0.28 to 0.40 mm) below the face of the front shell as shown by dimension A in Fig. 9.16. Leave the unit for a minimum of 24 hours to allow the Loctite to set hard.

18 Refit the servo unit to the car as described in the previous Section.

To test the servo unit for correct operation after overhaul first start the engine and run for a period of two minutes and then switch off. Wait for ten minutes and apply the footbrake very carefully, listening to hear the rush of air into the servo unit. This will indicate that vacuum was retained and, therefore, the unit is operating correctly.

22 Fault diagnosis - braking system

Before diagnosing faults from the following chart, check that any braking irregularities are not caused by:

1. Uneven and incorrect tyre pressures.
2. Incorrect 'mix' of radial and crossply tyres.
3. Wear in the steering mechanism.
4. Defects in the suspension and dampers.
5. Misalignment of the bodyframe.

Symptom	Reason/s
Pedal travels a long way before the brakes operate	Brake shoes set too far from the drums (auto. adjusters seized).
Stopping ability poor, even though pedal pressure is firm	Linings, discs or drums badly worn or scored. One or more wheel hydraulic cylinders seized, resulting in some brake shoes not pressing against the drums (or pads against discs). Brake linings contaminated with oil. Wrong type of linings fitted (too hard). Brake shoes wrongly assembled. Servo unit not functioning.
Car veers to one side when the brakes are applied	Brake pads or linings on one side are contaminated with oil. Hydraulic wheel cylinder(s) on one side partially or fully seized. A mixture of lining materials fitted between sides. Brake discs not matched. Unequal wear between sides caused by partially seized wheel cylinders.
Pedal feels spongy when the brakes are applied	Air is present in the hydraulic system.
Pedal feels springy when the brakes are applied	Brake linings not bedded into the drums (after fitting new ones). Master cylinder or brake backplate mounting bolts loose. Severe wear in brake drums causing distortion when brakes are applied. Discs out of true.
Pedal travels right down with little or no resistance and brakes are virtually non-operative	Leak in hydraulic system resulting in lack of pressure for operating wheel cylinders. If no signs of leakage are apparent the master cylinder internal seals are failing to sustain pressure.
Binding, juddering, overheating	One or a combination of reasons given in the foregoing Sections.

Chapter 10 Electrical system

For modifications, and information applicable to later models, see Supplement at end of manual

Contents

Alternator - fault diagnosis and repair ... 10	Light and windscreen wiper (front) switches - removal and refitting ... 58
Alternator - general ... 6	
Alternator - removal and refitting ... 9	Luggage compartment lamp - removal and refitting ... 57
Alternator - routine maintenance ... 7	Parking lamp bulb - removal and refitting ... 24
Alternator - special procedures ... 8	Radios and tape players - fitting (general) ... 60
Alternator (Bosch) brushes - removal, inspection and refitting ... 12	Radios and tape players - suppression of interference (general) ... 61
Alternator (Femsa) brushes - removal, inspection and refitting ... 13	Rear lamp assembly - bulb renewal ... 27
Alternator (Lucas) brushes - removal, inspection and refitting ... 11	Rear lamp assembly - removal and refitting ... 26
Battery - charging ... 5	Rear window washer and wiper switches - removal and refitting ... 56
Battery - electrolyte replenishment ... 4	Reversing light switch - removal and refitting ... 59
Battery - maintenance and inspection ... 3	Speedometer cable - renewal ... 46
Battery - removal and refitting ... 2	Starter motor - general description ... 14
Clock (console mounted) - removal and refitting ... 42	Starter motor (Bosch pre-engaged) - dismantling, overhaul and reassembly ... 21
Door pillar switches - removal and refitting ... 50	Starter motor (inertia) - dismantling, overhaul and reassembly ... 17
Facia mounted heated rear window switch and windscreen washer switch (front) - removal and refitting ... 55	Starter motor (inertia) - removal and refitting ... 16
Facia mounted heated rear window warning lamp bulb - renewal ... 54	Starter motor (inertia) - testing on engine ... 15
Fault diagnosis - electrical system ... 62	Starter motor (Lucas pre-engaged) - dismantling, overhaul and reassembly ... 20
Flasher unit ... 45	Starter motor (pre-engaged) - removal and refitting ... 19
Front direction indicator assembly - removal and refitting ... 25	Starter motor (pre-engaged) - testing on engine ... 18
Fuses ... 40	Steering column lock - removal and refitting ... 49
General description ... 1	Steering column multi-function switch - removal and refitting ... 44
Handbrake warning light bulb - renewal ... 53	Windscreen washer nozzle (front) - removal and refitting ... 34
Hazard warning switch - removal and refitting ... 43	Windscreen washer nozzle (rear) - removal and refitting ... 31
Headlamp assembly - removal and refitting ... 22	Windscreen washer pump (front) - removal and refitting ... 33
Headlamp beam - alignment ... 23	Windscreen washer pump (rear) - removal and refitting ... 30
Horn - fault finding and rectification ... 39	Windscreen wiper arms and blades - removal and refitting ... 38
Ignition switch - removal and refitting ... 48	Windscreen wiper mechanism - fault diagnosis and rectification ... 35
Instrument cluster - removal and refitting ... 41	Windscreen wiper motor - dismantling and reassembly ... 37
Instrument illumination and warning lamp bulbs (general) - renewal ... 52	Windscreen wiper motor and linkage (front) - removal and refitting ... 32
Instrument voltage regulator - removal and refitting ... 47	Windscreen wiper motor and linkage (rear) - removal and refitting ... 29
Interior light - removal and refitting ... 51	Windscreen wiper motor bush - renewal ... 36
Licence plate (number plate) lamp assembly - removal and refitting ... 28	

Specifications

System type ...	12 volt, negative earth
Battery type ...	Lead acid, 12 volt
Capacity (amp hour) ...	38

Starter motor (Lucas manufacture)

Type ...	M35J	5M90	2M100
Number of brushes ...	4	4	4
Brush material ...	Carbon	Carbon	Carbon
Minimum brush length ...	0.374 in (9.5 mm)	0.354 in (9.0 mm)	0.374 in (9.5 mm)
Brush spring pressure ...	16.94 oz (480 g)	30 oz (850 g)	16.94 oz (480 g)
Commutator:			
Minimum diameter ...	—	1.34 in (34 mm)	—
Maximum out of round ...	—	0.003 in (0.075 mm)	—
Armature endfloat ...	0.004 to 0.012 in (0.1 to 0.3 mm)	0.004 to 0.012 in (0.1 to 0.3 mm)	0.004 to 0.012 in (0.1 to 0.3 mm)
Type of drive ...	Solenoid	Solenoid	Solenoid
Number of pinion gear teeth ...	10	10	10
Number of teeth on flywheel ...	135	135	135
Direction of rotation ...	Clockwise	Clockwise	Clockwise

Maximum wattage draw (44 AH battery)	2,600	2,400	2,500
Voltage	12	12	12
Maximum output - watts	690	820	880

Starter on test

Maximum wattage draw no load @ 12V	740	900	940
Maximum wattage draw locked @ 7 volt supply @ terminals	2,730	2,590	3,325
Wattage draw at 180 rpm 20°C 44 AH battery	1,090	1,100	1,270

Starter motor (Bosch manufacture)

Type	EF 0.7	GF 1.0	GF 1.2
Minimum brush length	0.4 in (10 mm)	0.4 in (10 mm)	—
Brush spring pressure	32 to 46 oz (900 to 1300 g)	32 to 46 oz (900 to 1300 g)	—
Number of brushes	4	4	4
Brush material	Carbon Y-31	Carbon Y26 x 28	Carbon
Commutator:			
Minimum diameter	1.291 in (32.8 mm)	1.291 in (32.8 mm)	—
Maximum out of round	0.012 in (0.3 mm)	0.012 in (0.3 mm)	—
Armature endfloat	0.004 to 0.012 in (0.1 to 0.3 mm)	0.004 to 0.012 in (0.1 to 0.3 mm)	—
Maximum power draw (on load)	2400 watts	2500 watts	—
Voltage	12V	12V	12V
Output (on load)	515 watts	515 watts	—
Maximum power draw (off load)	540 watts	648 watts	950 watts
Type of drive	Solenoid	Solenoid	Solenoid
Number of pinion gear teeth	10	10	9
Number of flywheel ring gear teeth	135	135	121
Direction of rotation	Clockwise	Clockwise	Clockwise

Alternator (Bosch manufacture)

Type	G1-28A	K1-35A	K1-55A
Output at 13.5V and 6000 rpm (nominal)	28 amp	35 amp	55 amp
Stator winding resistance per phase	0.2 to 0.21 ohms	0.13 to 0.137 ohms	0.01 to 0.017 ohms
Rotor winding resistance at 20°C (68°F)	4 to 4.4 ohms	4 to 4.4 ohms	4 to 4.4 ohms
Minimum protrusion of brushes in free position	0.197 in (5 mm)	0.197 in (5 mm)	0.197 in (5 mm)
Regulating voltage (model AD1) 4000 rpm, 3 to 7 amp load	13.7 to 14.5 volt	13.7 to 14.5 volt	13.7 to 14.5 volt

Alternator (Femsa manufacture)

Type	ALD 12-32 or ALD 12-33
Output at 13.5V and 6000 rpm (nominal)	32 amp
Stator winding resistance per phase	0.173 ± 0.01 ohms
Rotor winding resistance at 20°C (68°F)	5.0 ± 0.15 ohms
Minimum protrusion of brushes in free position	0.28 in (7 mm)
Regulating voltage (model GRK 12-16), 4000 rpm, 3 to 7 amp load	13.7 to 14.5 volt
Field relay closing voltage	2.0 to 2.8 volt

Alternator (Lucas manufacture)

Type	Lucas 15 ACR	Lucas 17 ACR
Output @ 13.5V and 6000 rpm (nominal)	28 amp	35 amp
Stator winding resistance per phase	0.198 ± 0.01 ohms	0.133 ± 0.007 ohms
Rotor winding resistance @ 20°C (68°F)	3.27 ohms ± 5%	3.201 ohms ± 5%
Minimum protrusion of brushes in free position	0.2 in (5 mm)	0.2 in (5 mm)
Regulating voltage (model 14TR), 4000 rpm, 3 to 7 amp load	14.2 to 14.6 volt	14.2 to 14.6 volt

Windscreen wipers

Type	Two speed electric, self-parking

Horn

Type	4.0 in (102 mm) beep or projector
Current draw	4.5 to 5.0 amp

Bulb chart

Headlamp	45/40W
Headlamp (halogen)	55W
Sidelights (front)	4W
Licence plate	4W
Tail light	5W
Reversing light	21W
Stop light	21W
Rear fog light	21W
Direction indicators (front and rear)	21W
Interior lights	6W or 10W
Instrument panel warning light	2W or 2.6W
Instrument panel illumination	2W or 1.3W

For lamps not listed consult your Ford dealer

Fuses

Main fusebox on engine compartment bulkhead on driver's side

Fuse and rating	Circuits protected
1-16 amp	Cigarette lighter, clock, interior light
2-8 amp	Licence plate lights, instrument panel illumination
3-8 amp	RH tail and sidelights
4-8 amp	LH tail and sidelights
5-16 amp	Horn, blower motor
6-16 amp	Wiper motor, reversing lights
7-8 amp	Direction indicators, stoplights, instrument cluster

Fuses in dipper relay housing

Fuse and rating	Circuits protected
8-16 amp	LH dipped headlamp
9-16 amp	RH dipped headlamp
10-16 amp	RH main beam
10-16 amp	LH main beam

Fuses mounted under facia

Fuse and rating	Circuits protected
12-8 amp	Within relay for heated rear screen
13-2 amp	Radio circuit (medium-slow blow)
14-8 amp	Within relay for driving lamps (RPO)
15-8 amp	Within relay for fog lamps (RPO)

Torque wrench settings

	lb f ft	kg f m
Alternator pulley nut	25 to 29	3.5 to 4.0
Alternator mounting bolts	15 to 18	2.1 to 2.5
Alternator mounting bracket	20 to 25	2.8 to 3.5
Starter motor retaining bolts	20 to 25	2.8 to 3.5

1 General description

The major components of the 12 volt negative earth system comprise a battery, an alternator (driven from the crankshaft pulley), and a starter motor.

The battery supplies a regulated amount of current for the ignition, lighting and other electrical circuits and provides a reserve of power when the current consumed by the electrical equipment exceeds that being produced by the alternator.

The alternator has its own regulator which ensures a high output if the battery is in a low state of charge and the demand from the electrical equipment is high, and a low output if the battery is fully charged and there is little demand for the electrical equipment.

When fitting electrical accessories to cars with a negative earth system, it is important, if they contain silicone diodes or transistors, that they are connected correctly; otherwise serious damage may result to the components concerned. Items such as radios, tape players, electric ignition systems, electric tachometer, automatic dipping etc, should all be checked for correct polarity.

It is important that the battery negative lead is always disconnected if the battery is to be boost charged, and also if body repairs are to be carried out using electrical welding equipment. The alternator must be disconnected, otherwise serious damage can be caused. Whenever the battery has to be disconnected it must always be reconnected with the negative terminal earthed.

2 Battery - removal and refitting

1 The battery is on a carrier fitted to the left-hand wing valance of the engine compartment. It should be removed once every three months for cleaning and testing. Disconnect the negative and then the positive leads from the battery terminals by undoing and removing the plated nuts and bolts.

2 Unscrew and remove the bolt, and plain washer that secures the battery clamp plate to the carrier. Lift away the clamp plate. Carefully lift the battery from its carrier holding it vertically to ensure that none of the electrolyte is spilled.

3 Refitting is a direct reversal of this procedure. **Note:** Refit the positive lead before the negative lead and smear the terminals with petroleum jelly to prevent corrosion. **Never** use ordinary grease.

3 Battery - maintenance and inspection

1 Normal weekly battery maintenance consists of checking the electrolyte level of each cell to ensure that the separators are covered by ¼ in (6 mm) of electrolyte. If the level has fallen top up the battery using distilled water only. Do not overfill. If a battery is overfilled or any electrolyte spilled, immediately wipe away and neutralize as electrolyte attacks and corrodes any metal it comes into contact with very rapidly.

2 If the battery has the Auto-fil device fitted, a special topping up sequence is required. The white balls in the Auto-fil battery are part of the automatic topping up device which ensures correct electrolyte level. The vent chamber should remain in position at all times except when topping up or taking specific gravity readings. If the electrolyte level in any of the cells is below the bottom of the filling tube top up as follows:

   a) *Lift off the vent chamber cover.*
   b) *With the battery level, pour distilled water into the trough until all the filling tubes and trough are full.*
   c) *Immediately replace the cover to allow the water in the trough and tubes to flow into the cells. Each cell will automatically receive the correct amount of water.*

3 As well as keeping the terminals clean and covered with petroleum jelly, the top of the battery, and especially the top of the cells, should be kept clean and dry. This helps prevent corrosion and ensures that the battery does not become partially discharged by leakage through dampness and dirt.

4 Once every three months remove the battery and inspect the battery securing bolts, the battery clamp plate, tray, and battery leads for corrosion (white fluffy deposits on the metal which are brittle to touch). If any corrosion is found, clean off the deposits with ammonia and paint over the clean metal with an anti-rust/anti-acid paint.

5 At the same time inspect the battery case for cracks. If a crack is found, clean and plug it with one of the proprietary compounds marketed for this purpose. If leakage through the crack has been excessive then it will be necessary to refill the appropriate cell with fresh electrolyte as detailed later. Cracks are frequently caused to the top of the battery case by pouring in distilled water in the middle of winter *after* instead of *before* a run. This gives the water no chance to dilute the electrolyte and so the former freezes and splits the battery

case.
6 If topping up the battery becomes excessive and the case has been inspected for cracks that could cause leakage, but none are found, the battery is being overcharged and the voltage regulator will have to be checked and reset.
7 With the battery on the bench at the three monthly interval check, measure the specific gravity with a hydrometer to determine the state of charge and condition of the electrolyte. There should be very little variation between the different cells and if a variation in excess of 0.025 is present it will be due to either:

a) *Loss of electrolyte from the battery at sometime caused by spillage or a leak resulting in a drop in the specific gravity of the electrolyte, when the deficiency was replaced with distilled water instead of fresh electrolyte.*

b) *An internal short circuit caused by buckling of the plates or a similar malady pointing to the likelihood of total battery failure in the near future.*

8 The specific gravity of the electrolyte for fully charged conditions at the electrolyte temperature indicated, is listed in Table A. The specific gravity of a fully discharged battery at different temperatures of the electrolyte is given in Table B.

Table A
Specific Gravity - Battery Fully Charged
1.268 at 100°F or 38°C electrolyte temperature
1.272 at 90°F or 32°C electrolyte temperature
1.276 at 80°F or 27°C electrolyte temperature
1.280 at 70°F or 21°C electrolyte temperature
1.284 at 60°F or 16°C electrolyte temperature
1.288 at 50°F or 10°C electrolyte temperature
1.292 at 40°F or 4°C electrolyte temperature
1.296 at 30°F or -1.5°C electrolyte temperature

Table B
Specific Gravity - Battery Fully Discharged
1.098 at 100°F or 38°C electrolyte temperature
1.102 at 90°F or 32°C electrolyte temperature
1.106 at 80°F or 27°C electrolyte temperature
1.110 at 70°F or 21°C electrolyte temperature
1.114 at 60°F or 16°C electrolyte temperature
1.118 at 50°F or 10°C electrolyte temperature
1.122 at 40°F or 4°C electrolyte temperature
1.126 at 30°F or -1.5°C electrolyte temperature

4 Battery - electrolyte replenishment

1 If the battery is in a fully charged state and one of the cells maintains a specific gravity reading which is 0.025 or more lower than the others, and a check of each cell has been made with a voltmeter to check for short circuits (a four to seven second test should give a steady reading of between 12 to 18 volts) then it is likely that electrolyte has been lost from the cell with the low reading.
2 Top up the cell with a solution of 1 part sulphuric acid to 2.5 parts of water. If the cell is already fully topped-up draw some electrolyte out of it with an hydrometer.
3 When mixing the sulphuric acid and water **never add water to sulphuric acid** - always pour the acid slowly onto the water in a glass container. **If water is added to sulphuric acid it will explode.**
4 Continue to top up the cell with the freshly made electrolyte and then recharge the battery and check the hydrometer readings.

5 Battery - charging

1 In winter time, when heavy demand is placed upon the battery, such as starting from cold, and most of the electrical equipment is continually in use, it is a good idea to occasionally have the battery fully charged from an external source at the rate of 3.5 to 4 amps.
2 Continue to charge the battery at this rate until no further rise in specific gravity is noted over a four hour period.
3 Alternatively, a trickle charger charging at the rate of 1.5 amps can be safely used overnight.
4 Specially rapid 'boost' charges which are claimed to restore the power of the battery in 1 to 2 hours are not recommended as they can cause serious damage to the battery plates through over-heating.
5 While charging the battery, note that the temperature of the electrolyte should never exceed 100°F (37.8°C).

6 Alternator - general

The alternator may be of Lucas, Femsa or Bosch manufacture according to the vehicle and production source (Fig. 10.1).
The main advantage of the alternator over its predecessor, the dynamo, lies in its ability to provide a high charge at low revolutions. Driving slowly in heavy traffic with a dynamo invariably means no charge is reaching the battery. In similar conditions even with the wiper, heater, lights and perhaps radio switched on the alternator will ensure a charge reaches the battery.

7 Alternator - routine maintenance

1 The equipment has been designed for the minimum amount of maintenance in service, the only items subject to wear being the brushes and bearings.
2 Brushes should be examined after about 75,000 miles (120,000 km) and renewed if necessary. The bearings are prepacked with grease for life, and should not require further attention.
3 Check the fan belt every 3,000 miles (5,000 km) for correct adjustment which should be 0.5 in (13 mm) total movement at the centre of the longest run between pulleys.

8 Alternator - special procedures

Whenever the electrical system of the car is being attended to, and external means of starting the engine are used, there are certain precautions that must be taken otherwise serious and expensive damage can result.
1 Always make sure that the negative terminal of the battery is earthed. If the terminal connections are accidentally reversed or if the battery has been reverse charged the alternator diodes will be damaged.
2 The output terminal on the alternator marked 'BAT' or 'B+' must never be earthed but should always be connected directly to the positive terminal of the battery.
3 Whenever the alternator is to be removed or when disconnecting the terminals of the alternator circuit, always disconnect the earth terminal first.
4 The alternator must never be operated without the battery to alternator cable connected.
5 If the battery is to be charged by external means always disconnect both battery cables before the external charger is connected.
6 Should it be necessary to use a booster charger or booster battery

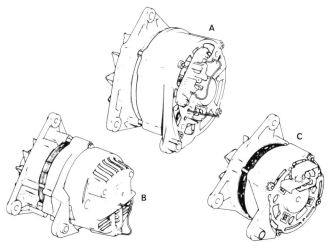

Fig. 10.1. The alternator recognition (Sec. 6)

A Bosch   B Lucas   C Femsa

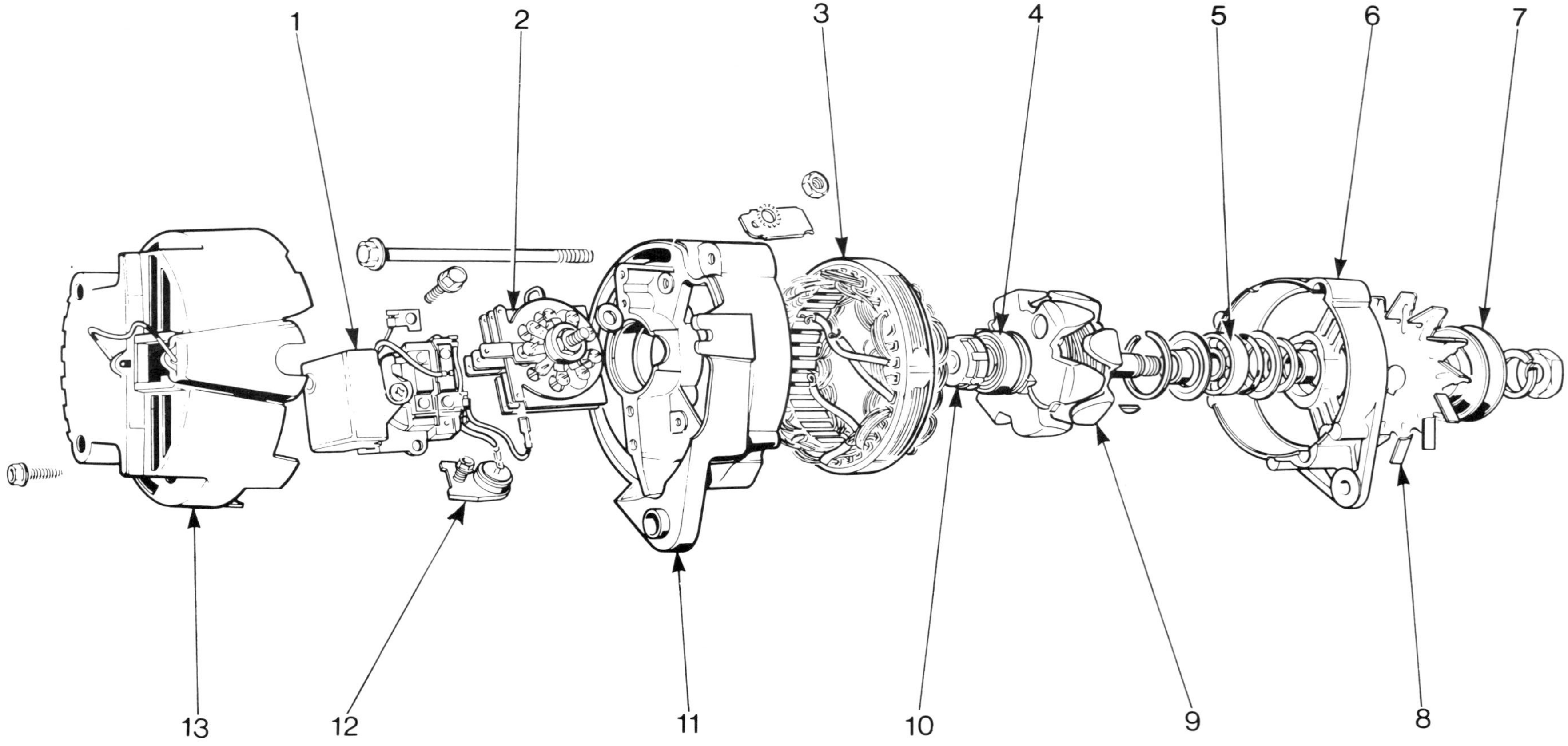

Fig. 10.2. The Lucas 15 ACR and 17 ACR alternator component parts (Sec. 11)

1 Regulator
2 Rectifier (diode) pack
3 Stator assembly
4 Slip ring end bearing
5 Drive end bearing
6 Drive end housing
7 Pulley
8 Fan
9 Rotor
10 Slip ring
11 Slip ring end housing
12 Surge protection diode
13 End cover

to start the engine always double check that the negative cable is connected to the negative terminal and the positive cable to positive terminal.

9 Alternator - removal and refitting

1 Disconnect the battery leads.
2 Note the terminal connections at the rear of the alternator and disconnect the plug or multi-pin connector.
3 Undo and remove the alternator adjustment arm bolt, slacken the alternator mounting bolts and push the alternator inward towards the engine. Lift away the fan belt from the pulley.
4 Remove the remaining two mounting bolts and carefully lift the alternator away from the car.
5 Take care not to knock or drop the alternator, otherwise this can cause irreparable damage.
6 Refitting the alternator is the reverse sequence to removal.
7 Adjust the fan belt so that it has a 0.5 in (13 mm) total movement at the centre of the longest run between pulleys.

10 Alternator - fault diagnosis and repair

Due to the specialist knowledge and equipment required to test or service an alternator it is recommended that if the performance is suspect the car be taken to an automobile electrician who will have the facilities for such work. Because of this recommendation, information is limited to the inspection and renewal of the brushes. Should the alternator not charge or the system be suspect the following points may be checked before seeking further assistance:

1 Check the fan belt tension, as described in Chapter 2, Section X.
2 Check the battery, as described in Section 3.
3 Check all electrical cable connections for cleanliness and security.

11 Alternator (Lucas) brushes - removal, inspection and refitting

1 Undo and remove the two screws and washers securing the end cover.
2 To inspect the brushes correctly the brush holder mountings should be removed complete by undoing the retaining bolts and disconnecting the 'Lucar' connection to the diode plates.
3 With the brush holder moulding removed and the brush assemblies still in position check that they protrude from the face of the moulding by at least 0.2 ins (5 mm). Also check that when depressed, the spring pressure is 7 to 10 ozs, when the end of the brush is flush with the face of the brush moulding. To be done with any accuracy this requires a push type spring gauge
4 Should either of the foregoing requirements not be fulfilled the brush assemblies should be renewed.
5 This can be done by simply removing the holding screws of each assembly and renewing them.
6 With the brush holder moulding removed the slip rings on the face end of the rotor are exposed. These can be cleaned with a petrol soaked cloth and any signs of burning may be removed very carefully with fine glass paper. On no account should any other abrasive be used or any attempt at machining be made.
7 When the brushes are refitted they should slide smoothly in their holders. Any sticking tendency may first be rectified by wiping with a petrol soaked cloth or, if this fails, by carefully polishing with a very fine file where any binding marks may appear.
8 Reassemble in the reverse order of dismantling. Ensure that leads which may have been connected to any of the screws are reconnected correctly. **Note:**

a) *If the charging system is suspect, first check the fan belt tension and condition - refer to Chapter 2, Section 9 for details.*
b) *Check the battery - refer to Section 3 for details.*
c) *With an alternator the ignition warning light control feed comes from the centre point of a pair of diodes in the alternator via a control unit similar in appearance to an indicator flasher unit. Should the warning light indicate lack of charge, check this unit and if suspect renew.*
d) *Should all the above prove satisfactory then proceed to check the alternator.*

12 Alternator (Bosch) brushes - removal, inspection and refitting

1 Undo and remove the two screws, spring and plain washers that secure the brush box to the rear of the end housing. Lift away the brush box (Figs. 10.5 and 7.7).
2 Check that the carbon brushes are able to slide smoothly in their guides without any sign of binding.
3 Measure the length of the brushes, and if they have worn down to 0.35 in (9 mm) or less, they must be renewed.
4 Hold the brush wire with a pair of pliers and unsolder it from the brush box. Lift away the two brushes and springs.
5 Insert the new springs and brushes and check to make sure that they are free to move in their guides. If they bond, lightly polish the terminals with a very fine file.
6 Solder the brush wire ends to the brush box making sure that a good connection is made.
7 Whenever new brushes are fitted new springs should also be fitted.
8 Refitting the brush box is the reverse sequence to removal.

13 Alternator (Femsa) brushes - removal, inspection and refitting

1 Disconnect the single wire from the brush box (Figs. 10.6 and 10.8).
2 Remove the retaining screw and withdraw the brush box.
3 Check that the carbon brushes are able to slide smoothly in their guides without any sign of binding.
4 Measure the amount by which the brushes protrude from the brush box. If this is less than 0.28 in (7 mm), renew the brushes.

Fig. 10.3. Brush box retaining screws - Lucas alternator (Sec. 11)

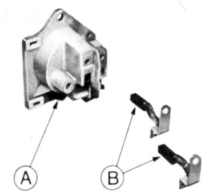

Fig. 10.4. Brush gear — Lucas alternator (Sec. 11)
A Brush box   B Brush assemblies

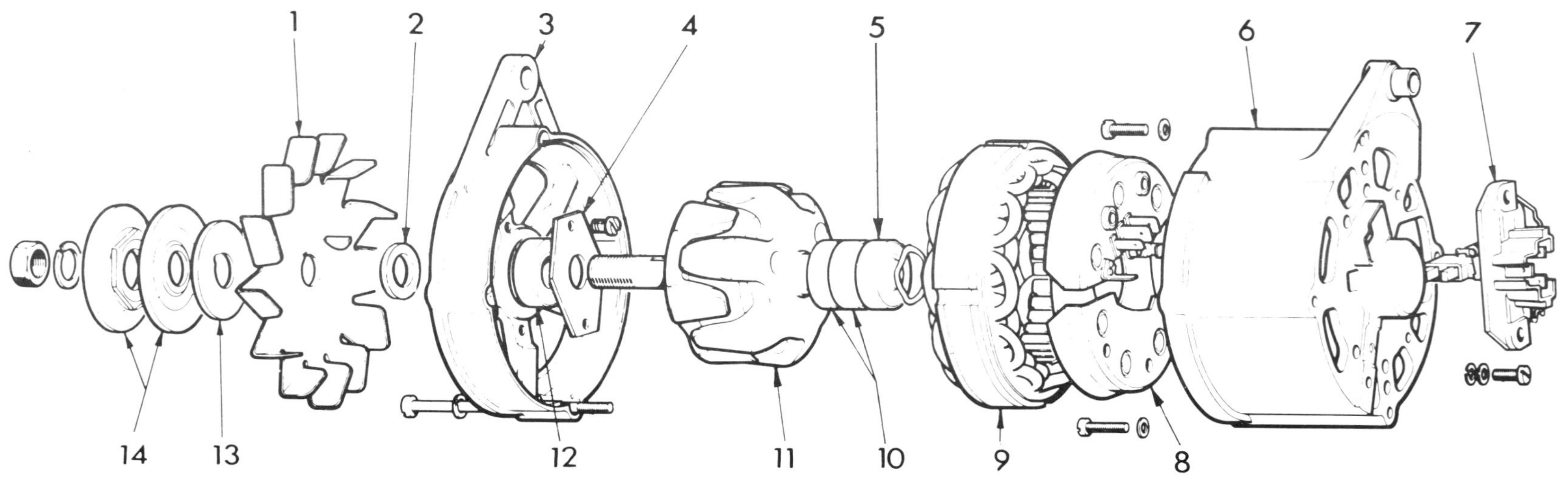

Fig. 10.5. Bosch alternator - exploded view (Sec. 12)

1 Fan
2 Spacer
3 Drive end housing
4 Thrust plate
5 Slip ring end bearing
6 Slip ring end housing
7 Brush box
8 Rectifier (diode) pack
9 Stator assembly
10 Slip rings
11 Rotor
12 Drive end bearing
13 Spacer
14 Pulley

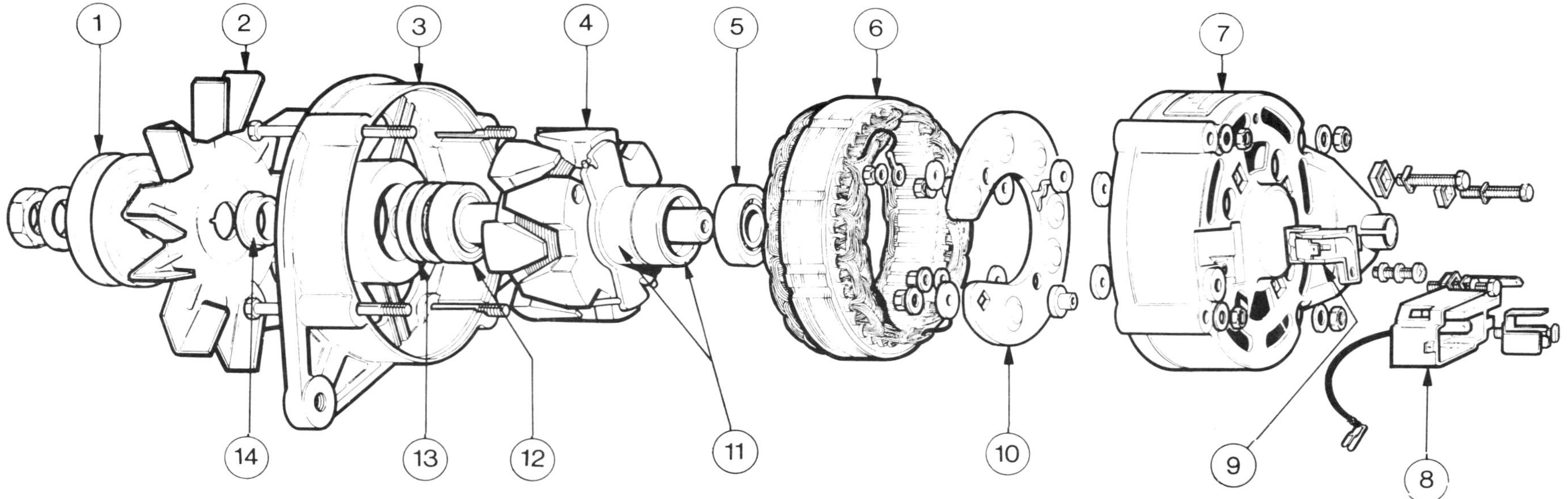

Fig. 10.6. Femsa alternator - exploded view (Sec. 13)

1 Pulley
2 Fan
3 Drive end housing
4 Rotor
5 Slip ring end bearing
6 Stator assembly
7 Slip ring end housing
8 Terminal block
9 Brush box
10 Rectifier (diode) pack
11 Slip rings
12 Drive end bearing
13 Thrust washers
14 Spacer

Chapter 10/Electrical system

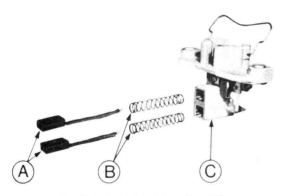

Fig. 10.7. Bosch brush box (Sec. 12)

A Brushes   B Springs   C Brush box

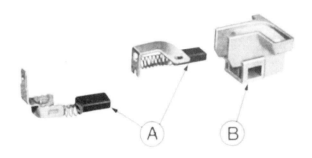

Fig. 10.8. Brush gear - Femsa alternator (Sec. 13)

A Brushes   B Brush box

5 Refitting the brush box is a straightforward reversal of the removal procedure.

14 Starter motor - general description

The starter motor fitted to the 1300 cc engine may be either of the inertia or pre-engaged type.

The pre-engaged type is recognisable by the solenoid assembly mounted on the motor body.

The principle of operation of the inertia type starter motor is as follows: When the ignition is switched on and the switch operated, current flows from the battery to the starter motor solenoid switch which causes it to become energised. Its internal plunger moves inwards and closes an internal switch so allowing full starting current to flow from the battery to the starter motor. This causes a powerful magnetic field to be induced into the field coils which causes the armature to rotate.

Mounted on helical spines is the drive pinion which, because of the sudden rotation of the armature, is thrown forward along the armature shaft and so into engagement with the flywheel ring gear. The engine crankshaft will then be rotated until the engine starts to operate on its own, and at this point, the drive pinion is thrown out of mesh with the flywheel ring gear.

The method of engagement on the pre-engaged starter differs considerably in that the drive pinion is brought into mesh with the starter ring gear before the main starter current is applied.

When the ignition is switched on, current flows from the battery to the solenoid which is mounted on the top of the starter motor. The plunger in the solenoid moves inwards so causing a centrally pivoted engagement lever to move in such a manner that the forked end pushes the drive pinion into mesh with the starter ring gear. When the solenoid reaches the end of its travel, it closes an internal contact and allows full starting current flow to the starter field coils. The armature is then able to rotate the crankshaft so starting the engine.

A special one way clutch is fitted to the starter drive pinion so that when the engine just fires and starts to operate on its own, it does not drive the starter motor.

15 Starter motor (inertia) - testing on engine

1 If the starter motor fails to operate, then check the condition of the battery by turning on the headlamps. If they glow brightly for several seconds and then gradually dim, the battery is in an uncharged condition.

2 If the headlamps continue to glow brightly and it is obvious that the battery is in good condition then check the tightness of the battery terminal to its connection on the body frame. Check the tightness of the connections at the relay switch and at the starter motor. Check the wiring with a voltmeter for breaks or shorts.

3 If the wiring is in order then check that the starter motor switch is operating. To do this, press the rubber covered button in the centre of the relay switch under the bonnet. If it is working, the starter motor will be heard to 'click', as it tries to rotate. Alternatively, check it with a voltmeter.

4 If the battery is fully charged, the wiring in order, and the switch working but the starter motor fails to operate, then it will have to be removed from the car for examination. Before this is done, however, ensure that the starter pinion has not jammed in mesh with the flywheel. Check by turning the square end of the armature shaft with a spanner. This will free the pinion if it is stuck in engagement with the flywheel teeth.

16 Starter motor (inertia) - removal and refitting

1 Disconnect the negative and then the positive terminals from the battery. Also disconnect the cable from the terminal on the starter motor end cover.

2 Undo and remove the nuts, bolts and spring washers which secure the starter motor to the clutch and the flywheel housing. Lift the starter motor away by manipulating the drive gear out from the ring gear area and then from the engine compartment.

3 Refitting is the reverse procedure to removal. Make sure that the starter motor cable, when secured in position by its terminal, does not touch any part of the body or power unit which could chafe the insulation.

17 Starter motor (inertia) - dismantling, overhaul and reassembly

1 With the starter motor placed on a bench or other suitable working area, loosen the screw on the cover band and slip the cover band off. With a piece of wire bent into the shape of a hook, lift back each of the brush springs in turn and check the movement of the brushes in their holders by pulling on the flexible connectors. If the brushes are so worn that their faces do not rest against the commutator, or if the ends of the brush leads are exposed on their working faces, they must be renewed.

2 If any of the brushes tend to stick in their holders then wash them with a petrol moistened cloth and, if necessary, lightly polish the sides of the brush with a very fine file until it moves quite freely in its holders.

3 If the surface of the commutator is dirty or blackened, clean it with a petrol dampened rag. Secure the starter motor in a vice and check it by connecting a heavy gauge cable between the starter motor terminal and a 12 volt battery.

4 Connect the cable from the other battery terminal to earth on the starter motor body. If the motor turns at high speed it is in good order.

5 If the starter motor still fails to function or if it is wished to renew the brushes then it is necessary to further dismantle the motor.

6 Lift the brush springs with the wire hook, and lift all four brushes out of their holders one at a time.

7 Remove the terminal nuts and washers from the terminal post on the commutator end bracket.

8 Unscrew the two through bolts which hold the end plates together and pull off the commutator end bracket. Also remove the driving end bracket which will come away complete with the armature.

9 At this stage, if the brushes are to be renewed, their flexible connectors must be unsoldered and the connectors of new brushes soldered in their place. Check that the new brushes move freely in their holders as detailed above. If cleaning the commutator with petrol fails to

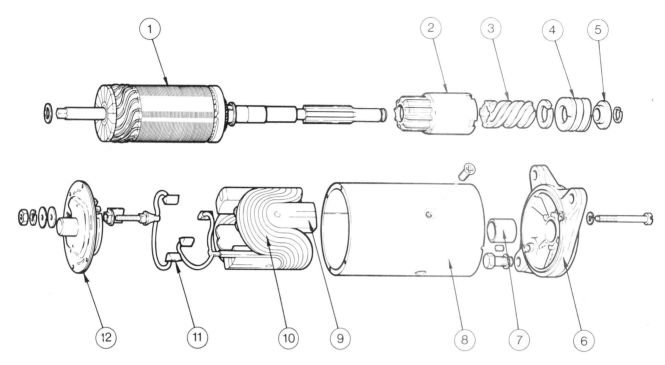

Fig. 10.9. Lucas inertia starter motor (Sec. 17)

1 Armature
2 Pinion
3 Bendix
4 Spring
5 Sleeve nut
6 Drive end housing
7 Drive end bush
8 Yoke
9 Pole segments
10 Field coils
11 Brushes
12 Commutator end housing

remove all the burnt areas and spots, then wrap a piece of glass paper round the commutator and rotate the armature.
10 If the commutator is very badly worn, remove the drive gear as detailed below. Then mount the armature in a lathe and with the lathe turning at high speed, take a very fine cut out of the commutator and finish the surface by polishing with glass paper. **Do not undercut the insulators between the commutator segments.**
11 With the starter motor dismantled, test the four field coils for an open circuit. Connect a 12 volt battery with a 12 volt bulb in one of the leads between the field terminal post and the tapping point of the field coils to which the brushes are connected. An open circuit is proved by the bulb not lighting.
12 If the bulb lights, it does not necessarily mean that the field coils are in order, as there is a possibility that one of the coils will be earthed to the starter yoke or pole shoes. To check this, remove the lead from the brush connector and place it against a clean portion of the starter yoke. If the bulb lights, the field coils are earthing. Renewal of the field coils calls for the use of a wheel operated screwdriver, a soldering iron, calking and riveting operations, and is beyond the scope of the majority of owners. The starter yoke should be taken to a reputable electrical engineering works for new field coils to be fitted. Alternatively purchase an exchange starter motor.
13 If the armature is damaged, this will be evident after visual inspection. Look for signs of burning, discolouration, and for conductors that have lifted away from the commutator.
14 With the starter motor stripped down, check the condition of the brushes. They should be renewed when they are sufficiently worn to allow visible side movement of the armature shaft.
15 The old bushes are simply driven out with a suitable drift and the new brushes inserted by the same method. As the brushes are of the phosphor bronze type it is essential that they are allowed to stand in engine oil for at least 24 hours before fitment. Alternatively soak in oil at 100°C (212°F) for 2 hours.
16 To dismantle the starter motor drive, first use a press to push the retainer clear of the circlip which can then be removed. Lift away the retainer and mainspring.
17 Slide the remaining parts with a rotary action off the armature shaft.
18 It is most important that the drive gear is completely free from oil, grease and dirt. With the drive gear removed, clean all parts thoroughly in paraffin. **Under no circumstances oil the drive components.** Lubrication of the drive components could easily cause the pinion to stick.
19 Reassembly of the starter motor drive is the reverse sequence to dismantling. Use a press to compress the spring and retainer sufficiently to allow a new circlip to be fitted in its groove on the shaft. Remove the drive from the press.
20 Reassembly of the starter motor is the reverse sequence to dismantling.

18 Starter motor (pre-engaged) - testing on engine

1 If the starter motor fails to operate, then check the condition of the battery by turning on the headlamps. If they glow brightly for several seconds and then gradually dim the battery is in an uncharged condition.
2 If the headlights continue to glow brightly and it is obvious that the battery is in good condition, then check the tightness of the battery wiring connections (and in particular the earth lead from the battery terminal to its connection on the body frame). If the positive terminal on the battery becomes hot when an attempt is made to operate the starter this is a sure sign of a poor connection on the battery terminal. To rectify, remove the terminal, clean the mating faces thoroughly and reconnect. Check the connections on the rear of the starter solenoid. Check the wiring with a voltmeter or test lamp for breaks or shorts.
3 Test the continuity of the solenoid windings by connecting a test lamp circuit comprising a 12 volt battery and low wattage bulb between the 'STA' terminal and the solenoid body. If the two windings are in order the lamp will light. Next connect the test lamp (fitted with a high wattage bulb) between the solenoid main terminals. Energise the solenoid by applying a 12 volt supply between the unmarked Lucar terminal and the solenoid body. The solenoid should be heard to operate and the test bulb light. This indicates full closure of the solenoid contacts.
4 If the battery is fully charged, the wiring in order, the starter/ignition switch working and the starter motor still fails to operate then it will have to be removed from the car for examination. Before this is done ensure that the starter motor pinion has not jammed in mesh with the flywheel by engaging a gear and rocking the car to and fro. This

Chapter 10/Electrical system

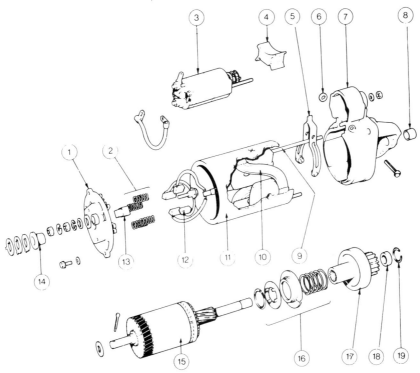

Fig. 10.10. Lucas pre-engaged starter motor (Sec. 20)

1 Commutator end housing	6 Pivot pin retaining clip	11 Yoke	15 Armature assembly
2 Brush springs	7 Drive end housing	12 Brushes	16 Drive plate and springs
3 Solenoid assembly	8 Drive end bush	13 Insulator	17 Pinion
4 Grommet	9 Through bolt	14 Commutator end bearing	18 Thrust collar
5 Pivot lever	10 Field coil		19 Ring

should free the pinion if it is stuck in mesh with the flywheel teeth.

19 Starter motor (pre-engaged) - removal and refitting

Removal is basically identical to that for the inertia type starter motor with the exception of the cables at the rear of the solenoid. Note these connections and then detach the cable terminal from the solenoid.

20 Starter motor (Lucas pre-engaged) - dismantling, overhaul and reassembly

1 Detach the heavy duty cable linking the solenoid 'STA' terminal to the starter motor terminal by undoing and removing the securing nuts and washers.
2 Undo and remove the two nuts and spring washers securing the solenoid to the drive end bracket.
3 Carefully withdraw the solenoid coil unit from the drive end bracket.
4 Lift off the solenoid plunger and return spring from the engagement lever.
5 Remove the rubber sealing block from the drive end bracket.
6 Remove the retaining ring (spire nut) from the engagement lever pivot pin and withdraw the pin.
7 Unscrew and remove the two drive end bracket securing nuts and spring washers and withdraw the bracket.
8 Lift away the engagement lever from the drive operating plate.
9 Extract the split pin from the end of the armature and remove the shim washers and thrust plate from the commutator end of the armature shaft.
10 Remove the armature together with its internal thrust washer.
11 Withdraw the thrust washer from the armature.
12 Undo and remove the two screws securing the commutator end bracket to the starter motor body yoke.
13 Carefully detach the end bracket from the yoke, at the same time disengaging the field brushes from the brush gear. Lift away the end bracket.

14 Move the thrust collar clear of the jump ring and then remove the jump ring. Withdraw the drive assembly from the armature shaft.
15 At this stage, if the brushes are to be renewed, their flexible connectors must be unsoldered and the connectors of the new brushes soldered in their place. Check that the new brushes move freely in their holders as detailed above. If cleaning the commutator with petrol fails to remove all the burnt areas and spots, then wrap a piece of glass paper around the commutator and rotate the armature.
16 If the commutator is very badly worn, remove the drive gear. Then mount the armature in a lathe and, with the lathe turning at high speed, take a very fine cut out of the commutator and finish the surface by polishing with glass paper. **Do not undercut the insulators between the commutator segments.**
17 With the starter motor dismantled, test the four field coils for an open circuit. Connect a 12 volt battery with a 12 volt bulb in one of the leads between the field terminal post and the tapping point of the field coils to which the brushes are connected. An open circuit is proved by the bulb not lighting.
18 If the bulb lights, it does not necessarily mean that the field coils are in order, as there is a possibility that one of the coils could be earthed to the starter yoke or pole shoes. To check this, remove the lead from the brush connector and place it against a clean portion of the starter yoke. If the bulb lights, the field coils are earthing. Renewal of the field coils calls for the use of a wheel operated screwdriver, a soldering iron, caulking and riveting operations, and is beyond the scope of the majority of owners. The starter yoke should be taken to a reputable electrical engineering works for new field coils to be fitted. Alternatively purchase an exchange Lucas starter motor.
19 If the armature is damaged this will be evident on inspection. Look for signs of burning, discolouration and for conductors that have lifted away from the commutator. Reassembly is a straightforward reversal of the dismantling procedure.
20 If a bearing is worn so allowing excessive side play of the armature shaft, the bearing bush must be renewed. Drift out the old bush with a piece of suitable diameter rod, preferably with a shoulder on it to stop the bush collapsing.
21 Soak a new bush in engine oil for 24 hours or, if time does not permit, heat in an oil bath at $100°C$ ($212°F$) for two hours prior to

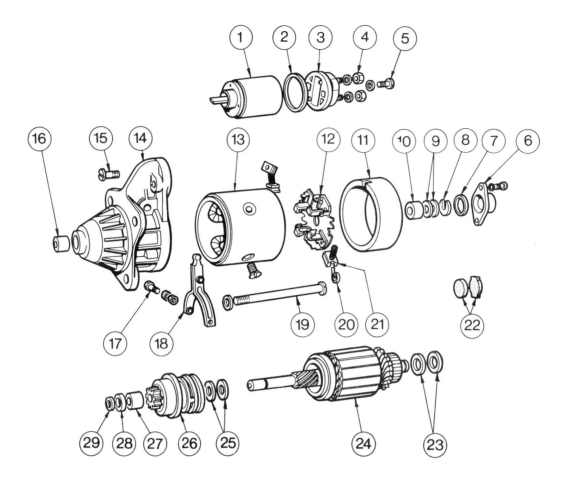

Fig. 10.11. Bosch pre-engaged starter motor (Sec. 21)

1 Solenoid assembly	8 'U' shoe	14 Drive end housing	22 Lubrication pads
2 Packing ring	9 Thrust washer	15 Screw	23 Thrust washers
3 Switch contacts and cover	10 Commutator end bearing	16 Bush	24 Armature assembly
4 Nut	11 Commutator end housing	17 Pivot pin	25 Packing rings
5 Screw	12 Brush plate	18 Pivot lever	26 Drive assembly
6 Cover	13 Yoke	19 Bolt	27 Bush
7 Washer		20 Brush spring	28 Stop ring
		21 Brush	29 Stop ring

fitting.

22 As a new bush must not be reamed after fitting, it must be pressed into position using a small mandrel of the same internal diameter as the bush and with a shoulder on it. Place the bush on the mandrel and press into position using a bench vice.

23 Using a test lamp and battery, test the continuity of the coil winding between terminal 'STA' and a good earth point on the solenoid body. If the lamp fails to light, the solenoid should be renewed.

24 To test the solenoid contacts for correct opening and closing, connect a 12 volt battery and a 60 watt test lamp between the main unmarked Lucar terminal and the 'STA' terminal. The lamp should not light.

25 Energise the solenoid with a separate 12 volt supply connected to the small unmarked Lucar terminal and a good earth on the solenoid body.

26 As the coil is energised the solenoid should be heard to operate and the test lamp should light with full brilliance.

27 The contacts may only be renewed as a set (ie. moving and fixed contacts). The fixed contacts are part of the moulded cover.

28 To fit a new set of contacts, first undo and remove the moulded cover securing screws.

29 Unsolder the coil connections from the cover terminals.

30 Lift away the cover and moving contact assembly.

31 Fit a new cover and moving contact assembly, soldering the connections to the cover terminals.

32 Refit the moulded cover securing screws.

33 Whilst the motor is apart, check the operation of the drive clutch. It must provide instantaneous take up of the drive in one direction and rotate easily and smoothly in the opposite direction.

34 Make sure that the drive moves freely on the armature shaft splines without binding or sticking.

35 Reassembly of the starter motor is the reverse sequence to dismantling. The following additional points should be noted:

a) When assembling the drive always use a new retaining ring (spire nut) to secure the engagement lever pivot pin.

b) Make sure that the internal thrust washer is fitted to the commutator end of the armature shaft before the armature is fitted.

c) Make sure that the thrust washers and plate are assembled in the correct order and are prevented from rotating separately, by engaging the collar pin with the locking piece on the thrust plate.

---

21 Starter motor (Bosch pre-engaged) - dismantling, overhaul and refitting

The procedure is similar to that described in the preceding Section but refer to the illustration for detail differences in component design (Fig. 10.11).

22 Headlamp assembly - removal and refitting

1 Disconnect the battery earth lead and remove the headlamp cover plate (photo).
2 Disengage the spring clip, then pull off the cap and multi-plug assembly (photo).
3 Remove the headlamp bulb retaine and bulb (photos).
4 Pull out the parking lamp bulb holder.
5 Remove the retaining screw and withdraw the headlamp assembly. If necessary, remove the adjusters and retaining clips from the lens and reflector assembly (photo).
6 Refitting is the reverse of the removal procedure, but it is recommended that beam alignment is checked and adjusted if necessary, as described in Section 23.

23 Headlamp beam - alignment

1 The procedure given in this Section is satisfactory for most practical purposes, although it is not intended to replace the alignment procedure used by many dealers and motor factors who would use beam setting equipment.
2 Refer to Fig. 10.13 which shows a beam setting chart for right-hand drive vehicles (for left-hand drive vehicles the chart is a mirror image).
3 Position the vehicle on flat, level ground, 33 ft (10 m) from a wall on which the aiming chart is to be fixed. A suitable chart can be drawn using white chalk on any convenient flat wall such as a garage wall or door.
4 Bounce the front of the vehicle to ensure that the suspension has settled and measure the height from the headlamp centre to the ground (H).
5 Mark the centre of the front and rear windows (outside if a heated rear screen is fitted) using a soft wax crayon or masking tape and position the car at right-angles to the chart so that:

   a) The vertical centre line and the window markings are exactly in line when viewed through the window and
   b) the horizontal line is at height 'H-X' above the ground.

6 Remove the headlamp cover plate and cover the right headlamp and switch on the main beam.
7 Adjust the horizontal alignment of the left-hand headlamp so that the intersection of the horizontal and angled light pattern coincides with the vertical line on the aiming chart (Fig. 10.14).
8 Adjust the vertical alignment so that the light/dark intersection of the beam pattern coincides with the dotted line on the aiming board.
9 Repeat the procedure for the left headlamp.
10 On completion, switch off the headlamps and refit the cover plates.

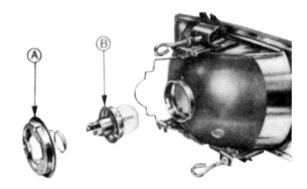

Fig. 10.12. Headlamp assembly (Sec. 22)

A Bulb retainer   B Bulb

22.1 Removing the headlamp cover plate

22.2 Headlamp cap and multi plug

22.3a Remove the bulb retainer ...

22.3b ... and bulb

22.5 The headlamp retaining screw

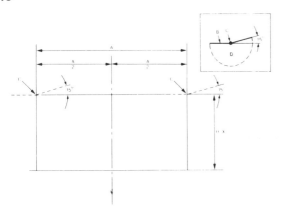

Fig. 10.13. Headlamp beam alignment chart (right-hand drive) (Sec. 23)

A Distance between headlamp centres
B Light/dark boundary
C Dipped beam centre
D Dipped beam pattern
H Height from ground to centre of headlamps
X 8 in (20 cm)

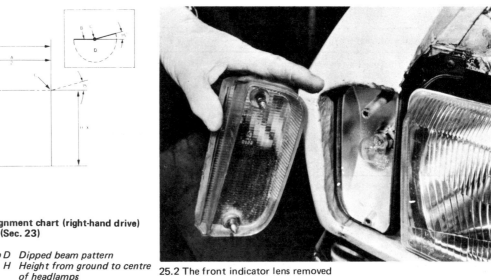

25.2 The front indicator lens removed

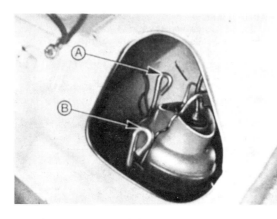

Fig. 10.14. Beam adjusting screws (Sec. 23)

A Horizontal   B Vertical

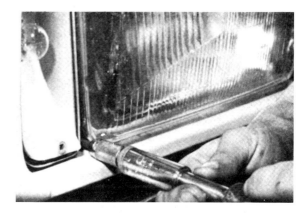

Fig. 10.15. Access to the direction indicator lamp lower screw (Sec. 25)

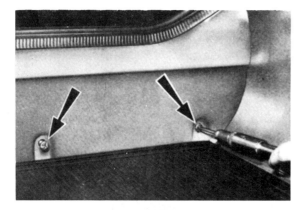

Fig. 10.16. Rear lamp assembly retaining screws (Sec. 26)

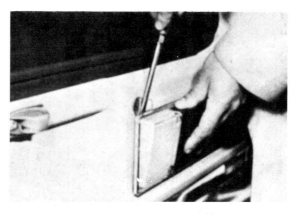

Fig. 10.17. Rear lamp assembly removal (Sec. 26)

Chapter 10/Electrical system

24 Parking lamp bulb - removal and refitting

1 Disconnect the battery earth lead and remove the headlamp cover plate.
2 Pull out the parking lamp bulb holder from the rear of the headlamp assembly.
3 Remove the parking lamp bulb.
4 Refitting is the reverse of the removal procedure.

25 Front direction indicator assembly - removal and refitting

1 Disconnect the battery earth lead and remove the headlamp cover plate.
2 Remove the indicator lens (2 screws) and remove the bulb (photo).
3 Remove the reflector. The lower retaining screw is accessible by inserting a screwdriver between the headlamp and reflector body. Disconnect the wiring to permit the assembly to be withdrawn.
4 Refitting is the reverse of the removal procedure.

26 Rear lamps assembly - removal and refitting

1 Disconnect the battery earth lead.
2 Open the tailgate, where applicable remove the tailgate trim panel, and remove the rear lamp retaining screws (Fig. 10.16).
3 Carefully lever out the rear lamp assembly and disconnect the wiring. Take care that the paintwork is not damaged during this operation (Fig. 10.17).
4 Clean off any caulking compound from the lamp body.
5 Installation is the reverse of the removal procedure, but to ensure a weather proof joint a caulking compound should be applied around the lamp body prior to its installation.

27 Rear lamp assembly - bulb renewal

1 Disconnect the battery earth lead.
2 Remove the four lens retaining screws and take off the lens (photo).
3 Remove and discard the bulbs as appropriate.
4 Refitting is the reverse of the removal procedure.

28 Licence plate (number plate) lamp assembly - removal and refitting

1 Disconnect the battery earth lead.
2 Open the tailgate, lift up the carpet and remove the spare wheel cover.
3 Remove the lamp lens (2 screws).
4 Disconnect the wiring and attach a length of cord to the lamp assembly lead to assist when installing so that the lead can be pulled through the body section. Remove the lamp body.
5 Refitting is the reverse of the removal procedure.
6 For access to the bulb only, remove the two crosshead screws and take off the lens (photo).

29 Windscreen wiper motor and linkage (rear) - removal and refitting

1 Disconnect the battery earth lead.
2 Remove the wiper arm and blade.
3 Open the tailgate, and remove the tailgate trim panel.
4 Disconnect the wiring at the wiper motor, noting the respective positions of the leads.
5 Remove the wiper spindle retaining nut and the three motor bracket retaining screws. Remove the motor and linkage assembly from the tailgate.
6 Remove the drive spindle nut and the three retaining bolts to detach the motor from the bracket.
7 Remove the circlip at the wiper spindle end and detach the linkage from the bracket.
8 Refitting is the reverse of the removal procedure, adjustment of motor bracket being made before the bolts are finally tightened.

30 Windscreen washer pump (rear) - removal and refitting

1 Disconnect the battery earth lead.
2 Open the tailgate and remove the spare wheel cover.
3 Remove the washer pipes and leads, noting their installed positions to prevent mix-up when refitting.
4 Remove the pump mounting screws and lift off the pump.
5 Refitting is the reverse of the removal procedure.

31 Windscreen washer nozzle (rear) - removal and refitting

1 Open the tailgate, remove the weather strip and pull down the headlining for access to the washer nozzle. Remove the nozzle (Fig. 10.20).
2 Refitting is the reverse of the removal procedure.

32 Windscreen wiper motor and linkage (front) - removal and refitting

1 Disconnect the battery earth lead.
2 Remove the windscreen wiper arm and blades. (Refer to Section

27.2 The rear lamp assembly lens removed

28.6 Access to the number plate lamp bulb

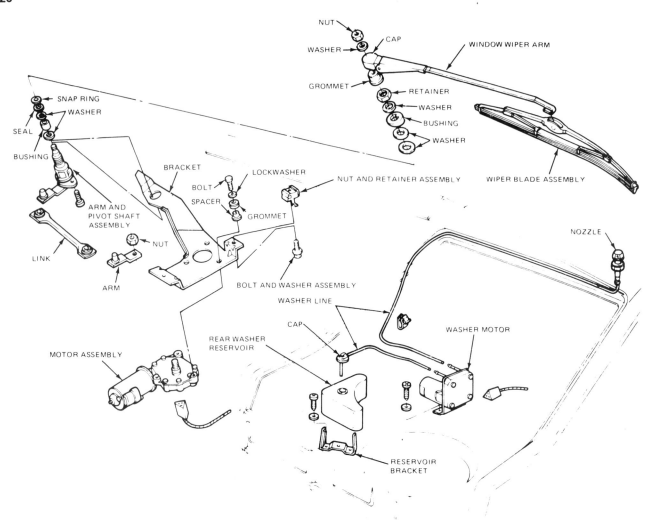

Fig. 10.18. Rear window washer and wiper assembly (Secs. 29 and 30)

Fig. 10.19. Adjustment point for the rear wiper assembly (Sec. 29)

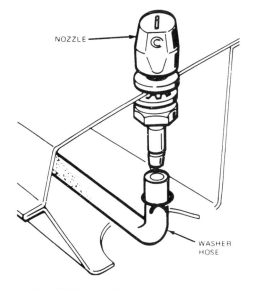

Fig. 10.20. Rear window washer nozzle (Sec. 31)

Chapter 10/Electrical system

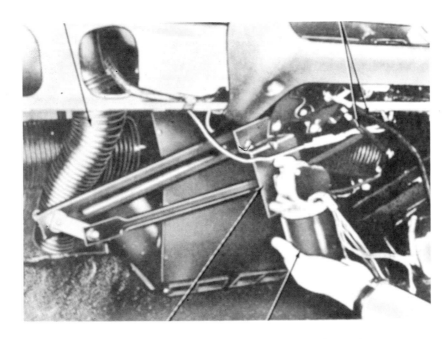

Fig. 10.21. Removing the windscreen wiper motor (Sec. 32)

38, if necessary).
3 Remove the air cleaner and disconnect the choke cable at the carburettor.
4 Remove the steering column shroud.
5 Remove the retaining screws, and pull the lower dash insulating panel and cover panel assembly clear of the dash panel.
6 Disconnect the cigar lighter wiring and withdraw the panel assembly, complete with choke cable, away from the vehicle.
7 Remove the instrument cluster bezel and the instrument cluster. Refer to Section 41, if necessary.
8 Remove the glovebox catch striker and glovebox assembly. Disconnect the light wiring.
9 Disconnect the cable from the heater controls.
10 Disconnect the driver's side demister tube connector from the heater box, and remove the connector and tube.
11 Disconnect and remove the driver's side face level vent tube.
12 Disconnect the wiring at the wiper motor and heater.
13 Remove the driver's side demister vent (1 screw).
14 Remove the wiper spindle retaining nuts, and the motor bracket retaining screw. Remove the motor and linkage from the vehicle.
15 If necessary, separate the motor from the linkage.
16 Refitting is the reverse of the removal procedure, but ensure that the heater control cable and the choke operating cable are correctly adjusted.

33 Windscreen washer pump (front) - removal and refitting

1 The front windscreen washers are operated from a facia mounted wash/wipe switch (see Section 56), and an integral pump and reservoir mounted at the front right-hand side of the engine compartment.
2 To remove the pump and reservoir, pull off the electrical connections, lift up the reservoir and disconnect the flexible pipe from the reservoir. The pump can be removed from the reservoir if necessary (photo).

34 Windscreen washer nozzles (front) - removal and refitting

1 Disconnect the battery earth lead.
2 Remove the retaining screws, withdraw the nozzles and disconnect the pipes.
3 Refitting is the reverse of the removal procedure.

33.2 Removing the front washer pump leads

35 Windscreen wiper mechanism - fault diagnosis and rectification

1 Should the windscreen wipers fail, or work very slowly, then first check the fuse. If this is in order, check the terminals on the motor for loose connections, and make sure the insulation of all the wiring is not cracked or broken thus causing a short circuit. If this is in order then check the current the motor is taking by connecting an ammeter in the circuit and turning on the wiper switch. Consumption should be between 2.3 to 3.1 amps.
2 If no current is passing through the motor, check that the switch is operating correctly.
3 If the wiper motor takes a very high current, check the wiper blades for freedom of movement. If this is satisfactory, check the gearbox cover and gear assembly for damage.
4 If the motor takes a very low current ensure that the battery is fully charged. Check the brush gear and ensure the brushes are bearing on the commutator. If not, check the brushes for freedom of movement and, if necessary, renew the tension springs. If the brushes are very worn they should be replaced with new ones. Check the armature by substitution if this unit is suspect.

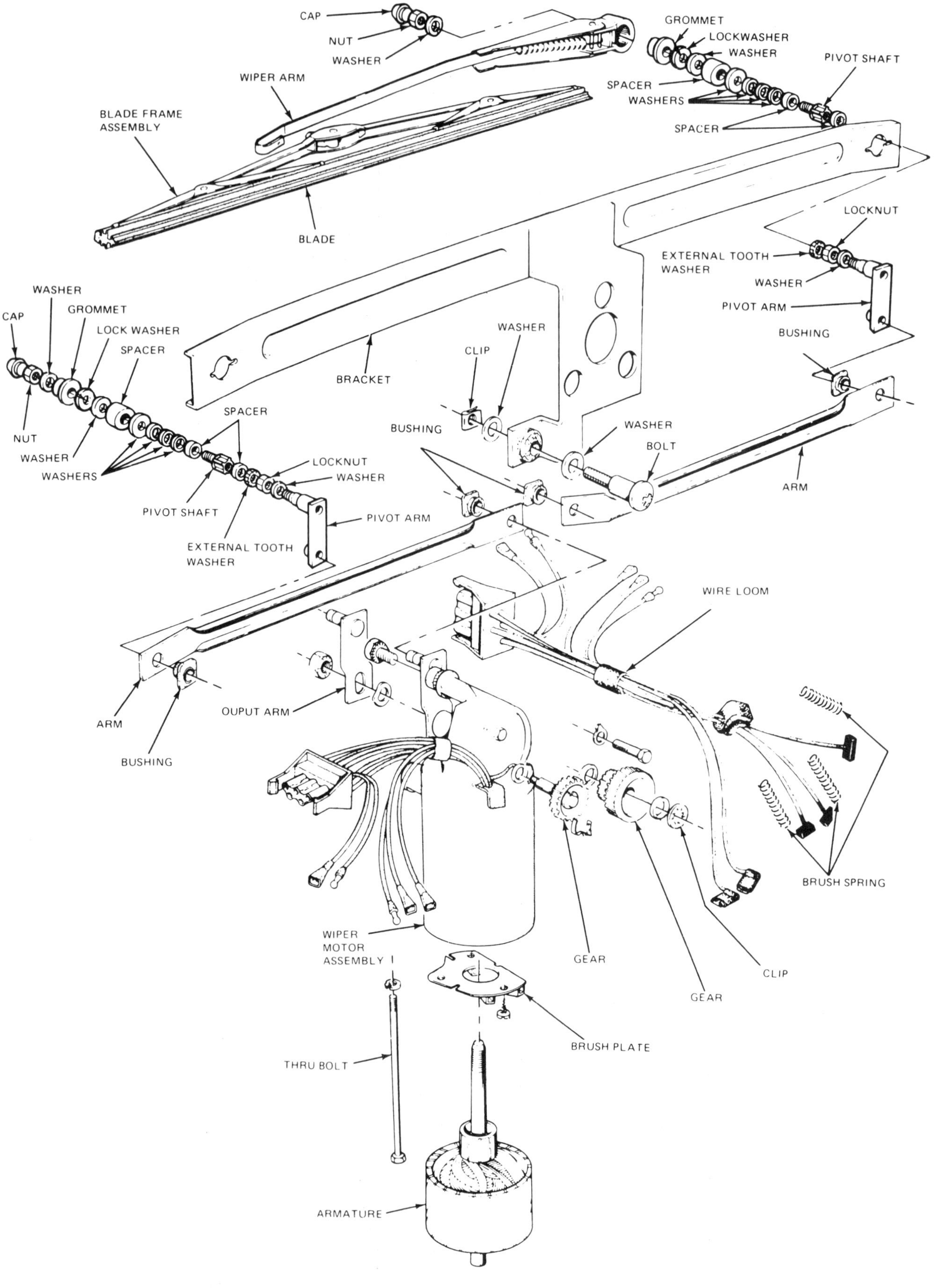

Fig. 10.22. Major parts of the windscreen wiper and linkage (Sec. 32)

Chapter 10/Electrical system

36 Windscreen wiper motor bush - renewal

1 Remove the two motor case/gear housing screws and withdraw the case and armature together.
2 Withdraw the brushes from the holders and remove the spring.
3 Remove the three brush mounting plate to wiper gear housing screws. Pull the wiring plug out of the side of the housing and remove the brush mounting plate.
4 Remove the screw and earth wire in the gear housing cover plate. Loosen the second screw and slide the cover plate away.
5 Disconnect the white/green and black/green leads from the terminals on the switch cover assembly, then remove the wiring assembly from the motor.
6 Disconnect the motor multi-pin connector from the harness and remove the motor feed wires and brushes.
7 Connect the replacement motor feed wire and brush assembly into the harness via the multi-pin connector.
8 Connect the black/green wire to the terminal marked 'black' and the white/green wire to the terminal marked 'green' on the switch cover assembly.
9 Slide the gear housing cover plate into position, ensuring that the wires are correctly positioned in the cut-out on the cover plate.

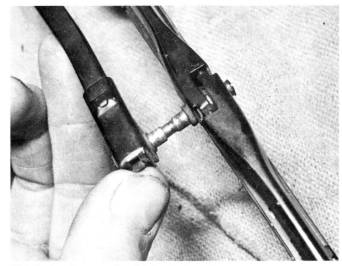

38.1 Spring clip type wiper blade attachment

37 Windscreen wiper motor - dismantling and reassembly

1 Separate the motor from the linkage.
2 Remove the brushes, wiring harness and brush mounting plate, referring to the previous Section as necessary.
3 Remove the remaining screw securing the gear housing cover plate and switch assembly. Remove the assembly.
4 Remove the spring steel armature stop from the gear housing.
5 Remove the spring clip and washer which secure the pinion gear; withdraw the gear and washers.
6 Remove the nut securing the motor output arm. Remove the arm, spring and flat washers.
7 Remove the output gear, the parking switch assembly and washer from the gear housing.
8 Reassembly is the reverse of the removal procedure, referring to the previous Section as necessary for the brush gear connections.

38 Windscreen wiper arms and blades - removal and refitting

1 To remove a wiper blade, raise the wiper arm away from the windscreen then, either slide the blade out of the hooked end of the arm or remove it from the spring clip in the centre of the blade. Refitting of the blade is straightforward (photo).
2 To remove a wiper arm, lift up the cap at the spindle end, remove the nut and carefully prise off the arm. When refitting, position the arm as necessary to obtain a satisfactory sweep on the windscreen (photo).

38.2 Wiper arm securing nut

39 Horn - fault finding and rectification

1 If the horn works badly or fails completely, check the wiring leading to the horn plug located on the body panel next to the horn itself. Also check that the plug is properly pushed home and is in a clean condition free from corrosion etc.
2 Check that the horn is secure on its mounting and that there is nothing lying on the horn body.
3 If the fault is not an external one, remove the horn cover and check the leads inside the horn. If they are sound, check the contact breaker contacts. If these are burnt or dirty clean them with a fine file and wipe all traces of dirt and dust away with a petrol moistened rag.

40 Fuses

1 If a fuse blows, always trace and rectify the cause before renewing it with one of the same rating.
2 The fuse block is located within the engine compartment on the side apron.

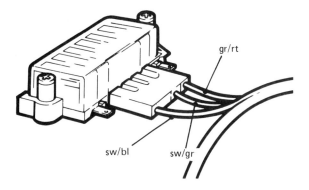

Fig. 10.23. The fuse block and connections (Sec. 40)

Cable colour code:
gr/rt   grey/red
sw/gr   black/grey
sw/bl   black/blue

3 The fuse ratings and circuits protected vary according to model and reference should be made to the Specifications Section at the beginning of this Chapter.

41 Instrument cluster - removal and refitting

1 Disconnect the battery earth lead.
2 Remove the steering column shroud. The bottom half is retained by two screws and the top half can then be pushed out.
3 Remove the retaining screws from the lower dash trim panel, ease the panel over the ignition switch and allow it to hang free.
4 Where applicable, pull out the two radio control knobs, remove three screws and disconnect the switch multi-plugs. Remove the facia panel.
5 Remove the four instrument cluster retaining screws and ease the cluster forwards. Disconnect the speedometer cable, oil pressure gauge feed pipe (where applicable) and the wiring loom multi-plug.
6 Refitting is the reverse of the removal procedure.

42 Clock (console mounted) - removal and refitting

Refer to the procedure given for removing the centre console in Chapter 12 where this item is listed.

43 Hazard warning switch - removal and refitting

1 Pull the switch from the lower dash trim panel and disconnect the wiring harness.
2 When refitting, connect the wiring harness and press the switch into the trim panel to retain it.

44 Steering column multi-function switch - removal and refitting

1 Disconnect the battery earth lead.
2 Remove the steering column shroud. The bottom half is retained by two screws and the top half can then be pulled out.
3 Remove the switch retaining screws, disconnect the multi-plug then detach the switch from the steering column.
4 Refitting is the reverse of the removal procedure.

45 Flasher unit

1 The flasher unit is mounted behind the instrument cluster and access can be gained to it after drawing the panel forward (see Section 41).
2 In the event of failure of a particular piece of equipment always check the connecting wiring, bulbs and fuses before assuming that it is the relay or flasher unit that is at fault. Take the relay or flasher unit to your dealer for testing or check the circuit by substituting a new component.

46 Speedometer cable - renewal

1 Chock the front wheels, jack up the rear of the car and support on firmly based stands.
2 Working under the car, carefully remove the snap-ring that secures the speedometer cable to the transmission. Detach the cable.
3 Now working in the engine compartment, remove the speedometer cable clip located on the engine bulkhead.
4 Ease the speedometer cable rubber grommet from the engine bulkhead.
5 Refer to Section 41 and move the instrument cluster by a sufficient amount to gain access to the rear of the speedometer.
6 Detach the cable from the rear of the speedometer.
7 Refitting is the reverse sequence to removal. For reliable operation, it is very important that there are no sharp bends in the cable run.

Fig. 10.24. Removing the instrument cluster (Sec. 41)

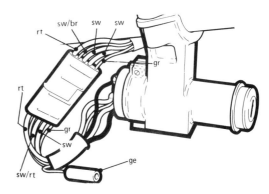

Fig. 10.25. The steering lock/ignition switch and cable connection (Secs. 44 and 48)

Cable colour code:

rt	red
sw/br	black/brown
sw	black
gr	grey
sw/rt	black/red
ge	yellow

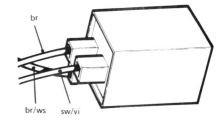

Fig. 10.26. The flasher unit (Sec. 45)

Cable colour code:

br	brown
sw/vi	black/violet

47 Instrument voltage regulator - removal and refitting

1 Remove the instrument panel cluster as described in Section 41.
2 Unscrew and remove the single screw that retains the instrument voltage regulator to the rear of the instrument panel and withdraw the regulator.
3 Refitting is a reversal of the removal procedure.

48 Ignition switch - removal and refitting

1 Disconnect the battery earth lead.
2 Remove the steering column shroud. The bottom half is retained by two screws and the top half can then be pulled out.
3 Set the ignition key to the 'O' position.
4 Note the location of the cables at the ignition switch and then detach the cables (Fig. 10.25).
5 Undo and remove the two screws that secure the ignition switch to the lock. Lift away the switch.
6 Refitting the ignition switch is the reverse of the removal procedure.

49 Steering column lock - removal and refitting

1 Disconnect the battery earth lead.
2 Remove the steering column shroud. The bottom half is retained by two screws and the top half can then be pushed out.
3 Undo and remove the two screws that secure the upper steering column support bracket.
4 Turn the column until it is possible to gain access to the headless bolts.
5 Note the location of the cables to the ignition switch terminals and lock body, and then detach the cables.
6 Using a suitable diameter drill, remove the headless bolts that clamp the lock to the steering column. Alternatively use a centre punch to rotate the bolts.
7 Lift away the lock assembly and clamp bracket.
8 Refitting the lock asssembly is the reverse sequence to removal. Make sure that the pawl enters the steering shaft. It will be necessary to use new shear bolts which must be tightened equally before the heads are separated from the shank.

50 Door pillar switches - removal and refitting

1 Disconnect the battery earth lead.
2 Prise the appropriate switch out of the door pillar, disconnect the lead and remove the switch.
3 Refitting is the reverse of the removal procedure.

51 Interior light - removal and refitting

1 To remove the interior light lens, switch, and/or body, carefully prise the lens away from the light body.
2 Refitting is the reversal of the removal procedure.

52 Instrument illumination and warning lamp bulbs (general) - renewal

1 Refer to Section 41 and move the instrument cluster by a sufficient amount to gain access to the bulb holder(s).
2 Extract the appropriate bulb from its holder.
3 Refitting is the reverse of the removal procedure.

53 Handbrake warning light bulb - renewal

1 Disconnect the battery earth lead.
2 Remove the steering column shroud. The bottom half is retained by two screws and the top half can then be pushed out.
3 Remove the ashtray.
4 Remove all the screws along the upper edge and glove compartment edge of the lower panel, and also those on the lower edge (outboard of the steering column) so that the lower panel can be pulled down and clear of the steering lock.
5 Reach up under the facia panel and apply sideways pressure to the bulb holder to release it from the instrument cluster.
6 Refitting is the reverse of the removal procedure, but ensure that the bulb holder electrical contacts are horizontal to mate with the printed circuit contacts.

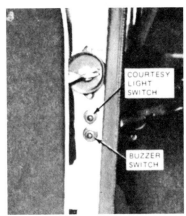

Fig. 10.27. Door pillar switches (Sec. 50)

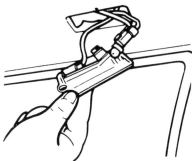

Fig. 10.28. Interior light removal (Sec. 51)

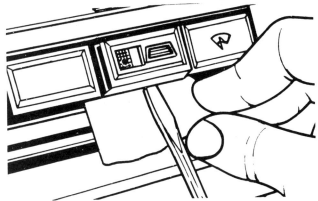

Fig. 10.29. Removing a facia mounted heated rear window warning light bulb, or a facia mounted windscreen washer switch (Sec. 54)

54 Facia mounted heated rear window warning light bulb - renewal

1 Using a piece of thick paper or a piece of card to prise against, use a screwdriver to prise out the switch assembly from the multi-plug.
2 Withdraw the bulb holder and remove the bulb.
3 Refitting is the reverse of the removal procedure.

55 Facia mounted heated rear window switch and windscreen washer switch (front) - removal and refitting

1 Follow the procedure given in the previous Section for warning light bulb renewal.

56 Rear window washer and wiper switch - removal and refitting

1 Remove the lower dash trim panel, as described in Section 41.
2 Disconnect the switch leads then press the switch(es) out of the trim panel.
3 Refitting is the reverse of the removal procedure.

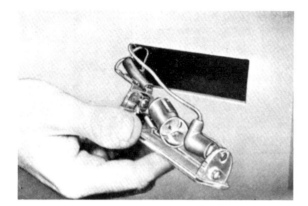

Fig. 10.30. Removing the luggage compartment lamp (Sec. 57)

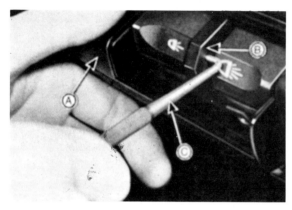

Fig. 10.31. Light and windscreen wiper (front) switch removal (Sec. 58)

A Lower panel   B Centre web   C Cranked tool

59.1 The reversing light switch location in the gearbox extension

57 Luggage compartment lamp - removal and refitting

1 Disconnect the battery earth lead.
2 Open the tailgate and pull out the lamp from the trim panel.
3 Note the relative positions of the electrical connections, then remove them from the lamp.
4 Refitting is the reverse of the removal procedure.

58 Light and windscreen wiper (front) switches - removal and refitting

1 Disconnect the battery earth lead.
2 Slacken the three screws on the lower panel.
3 Fully depress one of the switches of the pair to be removed, then insert a suitably cranked tool such as a piece of bent welding rod into the exposed hole in the switch centre web (Fig. 10.31).
4 Hold down the lower panel and gently pull out the switches.
5 When refitting, first connect the plug, then install the switch and tighten the panel screws.
6 Finally reconnect the battery earth lead.

59 Reverse light switch - removal and refitting

1 The reverse light switch is screwed into the rear end of the gearbox extension. To remove the switch, disconnect the leads and unscrew it from the extension. When refitting, apply a little non-setting gasket sealant on the screw threads and refit the leads (photo).

60 Radios and tape players - fitting (general)

A radio or tape player is an expensive item to buy and will only give its best performance if fitted properly. It is useless to expect concert hall performance from a unit that is suspended from the dash panel on string with its speaker resting on the back seat or parcel shelf! If you do not wish to do the installation yourself there are many in-car entertainment specialists who can do the fitting for you.

Make sure the unit purchased is of the same polarity as the car, and ensure that units with adjustable polarity are correctly set before commencing installation.

It is difficult to give specific information with regard to fitting, as final positioning of the radio/tape player, speakers and aerial is entirely a matter of personal preference. However, the following paragraphs give guidelines to follow, which are relevant to all installations.

Radios

Most radios are a standardised size of 7 inches wide, by 2 inches deep - this ensures that they will fit into the radio aperture provided in most cars. If your car does not have such an aperture, then the radio must be fitted in a suitable position either in, or beneath, the dashpanel. Alternatively, a special console can be purchased which will fit between the dashpanel and the floor, or on the transmission tunnel. These consoles can also be used for additional switches and instrumentation if required. Where no radio aperture is provided, the following points should be borne in mind before deciding exactly where to fit the unit:

a) *The unit must be within easy reach of the driver wearing a seat belt.*
b) *The unit must not be mounted in close proximity to an electric tachometer, the ignition switch and its wiring, or the flasher unit and associated wiring.*
c) *The unit must be mounted within reach of the aerial lead, and in such a place that the aerial will not have to be routed near the components detailed in the preceding paragraph 'b'..*
d) *The unit should not be positioned in a place where it might cause injury to the car occupants in an accident, for instance, under the dashpanel above the driver's or passenger's legs.*
e) *The unit must be fitted really securely.*

Some radios will have mounting brackets provided together with instructions: others will need to be fitted using drilled and slotted metal strips, bent to form mounting brackets - these strips are available from most accessory shops. The unit must be properly earthed by fitting a separate earthing lead between the casing of the radio and the vehicle frame.

Use the radio manufacturers' instructions when wiring the radio into the vehicle's electrical system. If no instructions are available refer to the relevant wiring diagram to find the location of the radio 'feed' connection in the vehicle's wiring circuit. A 1-2 amp 'in-line' fuse must be fitted in the radio 'feed' wire - a choke may also be

Chapter 10/Electrical system

necessary (see next Section).

The type of aerial used, and its fitted position is a matter of personal preference. In general the taller the aerial, the better the reception. It is best to fit a fully retractable aerial - especially, if a mechanical car-wash is used or if you live in an area where cars tend to be vandalised. In this respect electric aerials which are raised and lowered automatically when switching the radio on or off are convenient, but are more likely to give trouble than the manual type.

When choosing a site for the aerial the following points should be considered:

a) The aerial lead should be as short as possible - this means that the aerial should be mounted at the front of the car.

b) The aerial must be mounted as far away from the distributor and HT leads as possible.

c) The part of the aerial which protrudes beneath the mounting point must not foul the roadwheels, or anything else.

d) If possible the aerial should be positioned so that the coaxial lead does not have to be routed through the engine compartment.

e) The plane of the panel on which the aerial is mounted should not be so steeply angled that the aerial cannot be mounted vertically (in relation to the 'end-on' aspect of the car). Most aerials have a small amount of adjustment available.

Having decided on a mounting position, a relatively large hole will have to be made in the panel. The exact size of the hole will depend upon the specific aerial being fitted, although, generally, the hole required is of ¾ in (19 mm) diameter. On metal bodied cars a 'tank-cutter' of the relevant diameter is the best tool to use for making the hole. This tools needs a small diameter pilot hole drilled through the panel, through which, the tool clamping bolt is inserted. On GRP bodied cars a 'hole-saw' is the best tool to use. Again, this tool will require the drilling of a small pilot hole. When the hole has been made the raw edges should be de-burred with a file and then painted, to prevent corrosion.

Fit the aerial according to the manufacturer's instructions. If the aerial is very tall, or if it protrudes beneath the mounting panel for a considerable distance it is a good idea to fit a stay between the aerial and the vehicle frame. This stay can be manufactured from the slotted and drilled metal strips previously mentioned. The stay should be securely screwed or bolted in place. For best reception it is advisable to fit an earth lead between the aerial and the vehicle frame - this is essential for GRP bodied cars.

It will probably be necessary to drill one or two holes through bodywork panels in order to feed the aerial lead into the interior of the car. Where this is the case, ensure that the holes are fitted with rubber grommets to protect the cable, and to stop possible entry of water.

Positioning and fitting of the speaker depends mainly on its type. Generally, the speaker is designed to fit directly into the aperture already provided in the car (usually in the shelf behind the rear seats, or in the top of the dashpanel). Where this is the case, fitting the speaker is just a matter of removing the protective grille from the aperture and screwing or bolting the speaker in place. Take great care not to damage the speaker diaphragm whilst doing this. It is a good idea to fit a 'gasket' between the speaker frame and the mounting panel, in order to prevent vibration - some speakers will already have such a gasket fitted.

If a 'pod' type speaker was supplied with the radio, the best acoustic results will normally be obtained by mounting it on the shelf behind the rear seat. The pod can be secured to the mounting panel with self-tapping screws.

When connecting a rear mounted speaker to the radio, the wires should be routed through the vehicle beneath the carpets or floor mats - preferably the middle, or along the side of the floorpan, where they will not be trodden on by passengers. Make the relevant connections as directed by the radio manufacturer.

By now you will have several yards of additional wiring in the car, use PVC tape to secure this wiring out of harm's way. Do not leave electrical leads dangling. Ensure that all new electrical connections are properly made (wires twisted together will not do) and completely secure.

The radio should now be working, but before you pack away your tools, it will be necessary to 'trim' the radio to the aerial. If specific instructions are not provided by the radio manufacturer, proceed as follows. Find a station with a low signal strength on the medium-wave band, slowly turn the trim screw of the radio in or out until the loudest reception of the selected station is obtained - the set is then trimmed to the aerial.

Tape players

Fitting instructions for both cartridge and cassette stereo tape players are the same and in general the same rules apply as when fitting a radio. Tape players are not usually prone to electrical interference like radios - although it can occur - so positioning is not so critical. If possible the player should be mounted on an 'even-keel'. Also, it must be possible for a driver wearing a seat belt to reach the unit in order to change or turn over tapes.

For the best results from speakers designed to be recessed into a panel, mount them so that the back of the speaker protrudes into an enclosed chamber within the car (eg. door interiors or the boot cavity).

To fit recessed type speakers in the front doors, first check that there is sufficient room to mount the speakers in each door without it fouling the latch or window winding mechanism. Hold the speaker against the skin of the door and draw a line around the periphery of the speaker. With the speaker removed, draw a second 'cutting' line within the first, to allow enough room for the entry of the speaker back, but at the same time providing a broad seat for the speaker flange. When you are sure that the 'cutting-line' is correct, drill a series of holes around its periphery. Pass a hacksaw blade through one of the holes and then cut through the metal between the holes until the centre section of the panel falls out.

De-burr the edges of the hole and then paint the raw metal to prevent corrosion. Cut a corresponding hole in the door trim panel - ensuring that it will be completely covered by the speaker grille. Now drill a hole in the door edge and a corresponding hole in the door surround. These holes are to feed the speaker leads through - so fit grommets. Pass the speaker leads through the door trim, door skin and out through the holes in the side of the door and door surround. Refit the door trim panel and then secure the speaker to the door using self-tapping screws. **Note:** If the speaker is fitted with a shield to prevent water dripping on it, ensure that this shield is at the top.

Pod type speakers can be fastened to the shelf behind the rear seat, or anywhere else offering a corresponding mounting point on each side of the car. If the pod speakers are mounted on each side of the shelf behind the rear seat, it is a good idea to drill several large diameter holes through to the boot cavity beneath each speaker - this will improve the sound reproduction. Pod speakers sometimes offer a better reproduction quality if they face the rear window - which then acts as a reflector - so it is worthwhile to do a little experimenting before finally fixing the speaker.

---

61 Radios and tape players - suppression of interference (general)

To eliminate buzzes and other unwanted noises costs very little and is not as difficult as sometimes thought. With a modicum of common sense and patience and following the instructions in the following paragraphs, interference can be virtually eliminated.

The first cause for concern is the generator. The noise this makes over the radio is like an electric mixer and the noise speeds up when you rev up (if you wish to prove the point, you can remove the drivebelt and try it). The remedy for this is simple; connect a 1 to 3 mfd capacitor between earth, probably the bolt that holds down the generator base, and the *large* terminal on the dynamo or alternator. This is most important for if you connect it to the small terminal, you will probably damage the generator permanently (see Fig. 10.32).

A second common cause of electrical interference is the ignition system. Here a 1 mfd capacitor must be connected between earth and the 'SW' or '+' terminal on the coil (see Fig. 10.33). This may stop the tick-tick-tick sound that comes over the speaker. Next comes the spark itself.

There are several ways of curing interference from the ignition HT system. One is to use carbon film HT leads but these have a tendency to 'snap' inside and you don't know then, why you are firing on only half your cylinders. So the second, and more successful method is to use resistive spark plug caps (see Fig. 10.34) of about 10,000 to 15,000 ohms resistance. If, due to lack of room, these cannot be used, an alternative is to use 'in-line' suppressions (Fig. 10.34) - if the interference is not too bad, you may get away with only one suppressor in the coil to distributor line. If the interference does continue (a

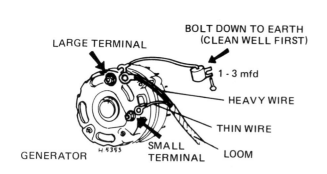

Fig. 10.32. The correct way to connect a capacitor to the generator

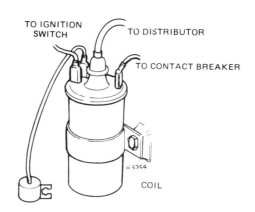

Fig. 10.33. The capacitor must be connected to the ignition switch side of the coil

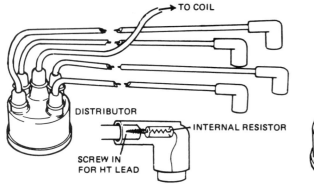

Resistive spark plug caps

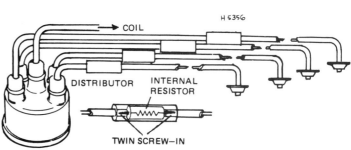

Fig. 10.34. Ignition HT lead suppressors

'In-line' suppressors

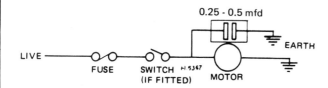

Fig. 10.35. Correct method of suppressing electric motors

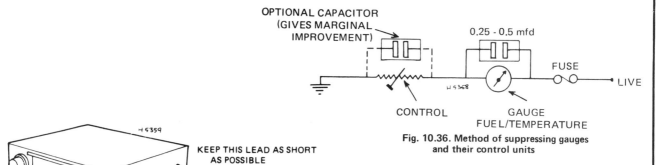

Fig. 10.36. Method of suppressing gauges and their control units

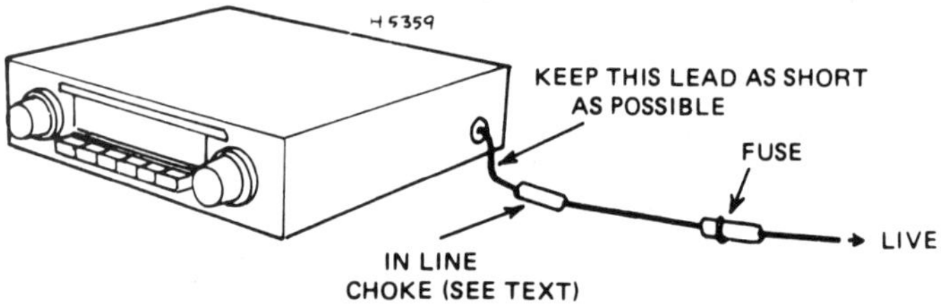

Fig. 10.37. An 'in-line' choke should be fitted into the live supply lead as close to the unit as possible

Chapter 10/Electrical system

'clacking' noise) then doctor all HT leads.

At this stage it is advisable to check that the radio is well earthed, also the aerial, and to see that the aerial plug is pushed well into the set and that the radio is properly trimmed (see preceding Section). In addition check that the wire which supplies the power to the set is as short as possible and does not wander all over the car. At this stage it is a good idea to check that the fuse is of the correct rating. For most sets this will be about 1 to 2 amps.

At this point the more usual causes of interference have been suppressed. If the problem still exists, a look at the causes of interference may help to pinpoint the component generating the stray electrical discharges.

The radio picks up electromagnetic waves in the air; now some are made by radio stations and other broadcasters and some, not wanted, are made by the car. The home made signals are produced by stray electrical discharges floating around the car. Common producers of these signals are electric motors; ie, the windscreen wipers, electric screen washers, electric window winders, heater fan or an electric aerial, if fitted. Other sources of interference are electrical fuel pumps, flashing turn signals, and instruments. The remedy for these cases is shown in Fig. 10.35 for an electric motor whose interference is not too bad and Fig. 10.36 for instrument suppression. Turn signals are not normally suppressed. In recent years, radio manufacturer's have included in the line (live) of the radio, in addition to the fuse, an 'in-line' choke. If your installation lacks one of these, put one in as shown in Fig. 10.37.

All the foregoing components are available from radio shops or accessory shops. For a transistor radio, a 2A choke should be adequate. If you have an electric clock fitted this should be suppressed by connecting a 0.5 mfd capacitor directly across it as shown for a motor in Fig. 10.35.

If, after all this, you are still experiencing radio interference, first assess how bad it is, for the human ear can filter out unobtrusive unwanted noises quite easily. But if you are still adamant about eradicating the noise, then continue.

As a first step, a few 'experts' seem to favour a screen between the radio and the engine. This is OK as far as it goes, literally! - for the whole set is screened and if interference can get past that then a small piece of aluminium is not going to stop it.

A more sensible way of screening is to discover if interference is coming down the wires. First, take the live lead; interference can get between the set and the choke (hence the reason for keeping the wires short). One remedy here is to screen the wire and this is done by buying screened wire and fitting that. The loudspeaker lead could be screened also to prevent 'pick-up' getting back to the radio - although this is unlikely.

Without doubt, the worst source of radio interference comes from the ignition HT leads, even if they have been suppressed. The ideal way of suppressing these is to slide screening tubes over the leads themselves. As this is impractical, we can place an aluminium shield over the majority of the lead areas. In a vee- or twin-cam engine, this is relatively easy but for a straight engine the results are not particularly good.

Now for the really impossible cases, here are a few tips to try out. Where metal comes into contact with metal, an electrical disturbance is caused which is why good clean connections are essential. To remove interference due to overlapping or butting panels you must bridge the join with a wide braided earth strap (like that from the frame to the engine/transmission). The most common moving parts that could create noise and should be strapped are, in order of importance:

a) *Silencer to frame.*
b) *Exhaust pipe to engine block and frame.*
c) *Air cleaner to frame.*
d) *Front and rear bumpers to frame.*
e) *Steering column to frame.*
f) *Hood and trunk lids to frame.*
g) *Hood frame to frame on soft tops.*

These faults are most pronounced when (1) the engine is idling, (2) labouring under load. Although the moving parts are already connected with nuts, bolts, etc, these do tend to rust and corrode, thus creating a high resistance interference source.

If you have a 'ragged' sounding pulse when mobile, this could be wheel or tyre static. This can be cured by buying some anti-static powder and sprinkling it liberally inside the tyres.

If the interference takes the shape of a high pitched screeching noise that changes its note when the car is in motion and only comes now and then, this could be related to the aerial, especially if it is of the telescopic or whip type. This source can be cured quite simply by pushing a small rubber ball on top of the aerial (yes, really) as this breaks the electric field before it can form; but it would be much better to buy yourself a new aerial of a reputable brand. If, on the other hand, you are getting a loud rushing sound every time you brake, then this is brake static. This effect is most prominent on hot dry days and is cured only by fitting a special kit, which is quite expensive.

In conclusion, it is pointed out that it is relatively easy, and therefore cheap to eliminate **95 per cent** of all noises, but to eliminate the final **5 per cent** is time and money consuming. It is up to the individual to decide if it is worth it. Please remember also, that you will not get concert hall performance from a cheap radio.

Finally, at the beginning of this Section are mentioned tape players; these are not usually affected by interference but in a very bad case, the best remedies are the first three suggestions plus using a 3 - 5 amp choke in the 'live' line and in incurable cases screen the live and speaker wires.

Note: If your car is fitted with electronic ignition, then it is not recommended that either the spark plug resistors or the ignition coil capacitor be fitted as these may damage the system. Most electronic ignition units have built-in suppression and should, therefore, not cause interference.

Fault diagnosis overleaf

62 Fault diagnosis - electrical system

Symptom	Reason/s
No voltage at starter motor	Battery discharged. Battery defective internally. Battery terminal leads loose or earth lead not securely attached to body. Loose or broken connections in starter motor circuit. Starter motor switch or solenoid faulty.
Voltage at starter motor: faulty motor	Starter motor pinion jammed in mesh with flywheel gear ring. Starter brushes badly worn, sticking, or brush wires loose. Commutator dirty, worn or burnt. Starter motor armature faulty. Field coils earthed.
Electrical defects	Battery in discharged condition. Starter brushes badly worn, sticking, or brush wires loose. Loose wires in starter motor circuit.
Dirt or oil on drive gear	Starter motor pinion sticking on the screwed sleeve.
Mechanical damage	Pinion or flywheel gear teeth broken or worn.
Lack of attention or mechanical damage	Pinion or flywheel gear teeth broken or worn. Starter drive main spring broken. Starter motor retaining bolts loose.
Wear or damage	Battery defective internally. Electrolyte level too low or electrolyte too weak due to leakage. Plate separators no longer fully effective. Battery plates severely sulphated.
Insufficient current flow to keep battery charged	Fan belt slipping. Battery terminal connections loose or corroded. Alternator not charging properly. Short in lighting circuit causing continual battery drain. Regulator unit not working correctly.
Alternator not charging *	Fan belt loose and slipping, or broken. Brushes worn, sticking, broken or dirty. Brush springs weak or broken.

If all appears to be well but the alternator is still not charging, take the car to an automobile electrician for checking of the alternator and regulator

Symptom	Reason/s
Battery will not hold charge for more than a few days	Battery defective internally. Electrolyte level too low or electrolyte too weak due to leakage. Plate separators no longer fully effective. Battery plates severely sulphated. Fan/alternator belt slipping. Battery terminal connections loose or corroded. Alternator not charging properly. Short in lighting circuit causing continual battery drain. Regulator unit not working correctly.
Ignition light fails to go out, battery runs flat in a few days	Fan belt loose and slipping or broken. Alternator faulty.

Failure of individual electrical equipment to function correctly is dealt with alphabetically, below

Symptom	Reason/s
Fuel gauge gives no reading	Fuel tank empty! Electric cable between tank sender unit and gauge earthed or loose. Fuel gauge case not earthed. Fuel gauge supply cable interrupted. Fuel gauge unit broken.
Fuel gauge registers full all the time	Electric cable between tank unit and gauge broken or disconnected.
Horn operates all the time	Horn push either earthed or stuck down. Horn cable to horn push earthed.
Horn fails to operate	Blown fuse. Cable or cable connection loose, broken or disconnected. Horn has an internal fault.

Chapter 10/Electrical system

Symptom	Reason/s
Horn emits intermittent or unsatisfactory noise	Cable connections loose. Horn incorrectly adjusted.
Lights do not come on	If engine not running, battery discharged. Light bulb filament burnt out or bulbs broken. Wire connections loose, disconnected or broken. Light switch shorting or otherwise faulty.
Lights come on but fade out	If engine not running battery discharged.
Lights give very poor illumination	Lamp glasses dirty. Reflector tarnished or dirty. Lamps badly out of adjustment. Incorrect bulb with too low wattage fitted. Existing bulbs old and badly discoloured. Electrical wiring too thin not allowing full current to pass.
Lights work erratically - flashing on and off, especially over bumps	Battery terminals or earth connections loose. Lights not earthing properly. Contacts in light switch faulty.
Wiper motor fails to work	Blown fuse. Wire connections loose, disconnected or broken. Brushes badly worn. Armature worn or faulty. Field coils faulty.
Wiper motor works very slowly and takes excessive current	Commutator dirty, greasy or burnt. Drive to spindles too bent or unlubricated. Drive spindle binding or damaged. Armature bearings dry or unaligned. Armature badly worn or faulty.
Wiper motor works slowly and takes little current	Brushes badly worn. Commutator dirty, greasy or burnt. Armature badly worn or faulty.
Wiper motor works but wiper blades remain static	Linkage disengaged or faulty. Drive spindle damaged or worn. Wiper motor gearbox parts badly worn.

The following codes are applicable to all circuit diagrams on pages 132 - 138

Wire codes

```
54 - 16   sw/gr-rt   2.5
                       └── Wire cross-section in mm². Unmarked wires have 0.75 mm² cross-section
              └──────── Wire colour code – secondary colours
              └──────── Wire colour code – main colour
   └──────────────────── Wire number
```

Wiring colour	Code	Wiring colour	Code
Blue	bl	Pink	rs
Brown	br	Red	rt
Yellow	ge	Black	sw
Grey	gr	Violet	vi
Green	gn	White	ws

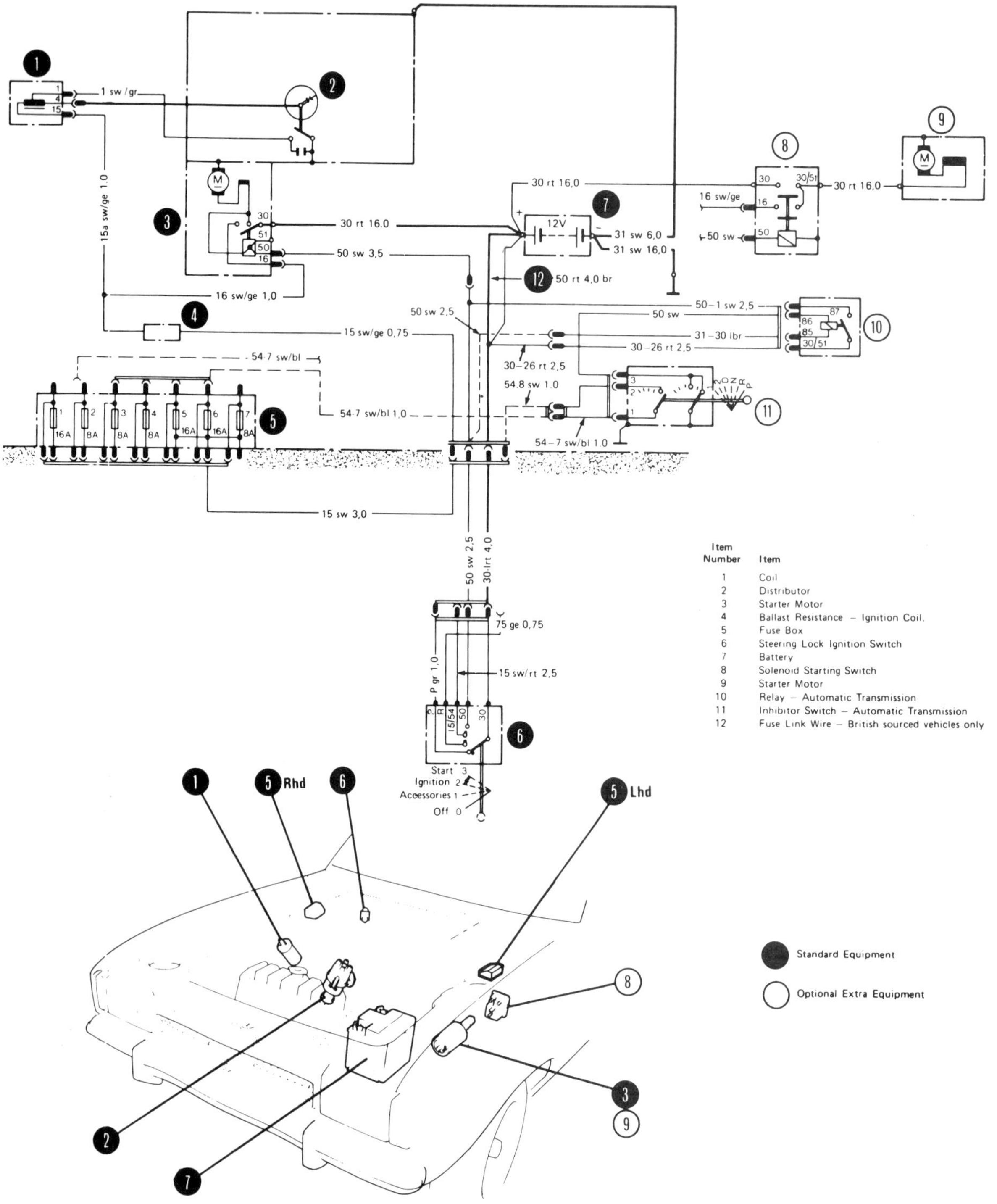

Item Number	Item
1	Coil
2	Distributor
3	Starter Motor
4	Ballast Resistance – Ignition Coil
5	Fuse Box
6	Steering Lock Ignition Switch
7	Battery
8	Solenoid Starting Switch
9	Starter Motor
10	Relay – Automatic Transmission
11	Inhibitor Switch – Automatic Transmission
12	Fuse Link Wire – British sourced vehicles only

● Standard Equipment
○ Optional Extra Equipment

Fig. 10.38. Starting and ignition circuits

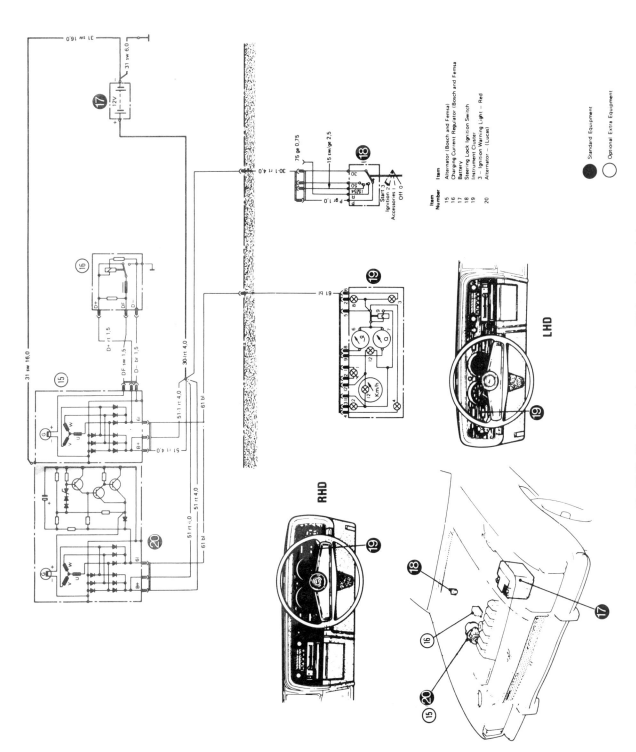

Fig. 10.39. Charging circuits (see page 132 for wiring codes)

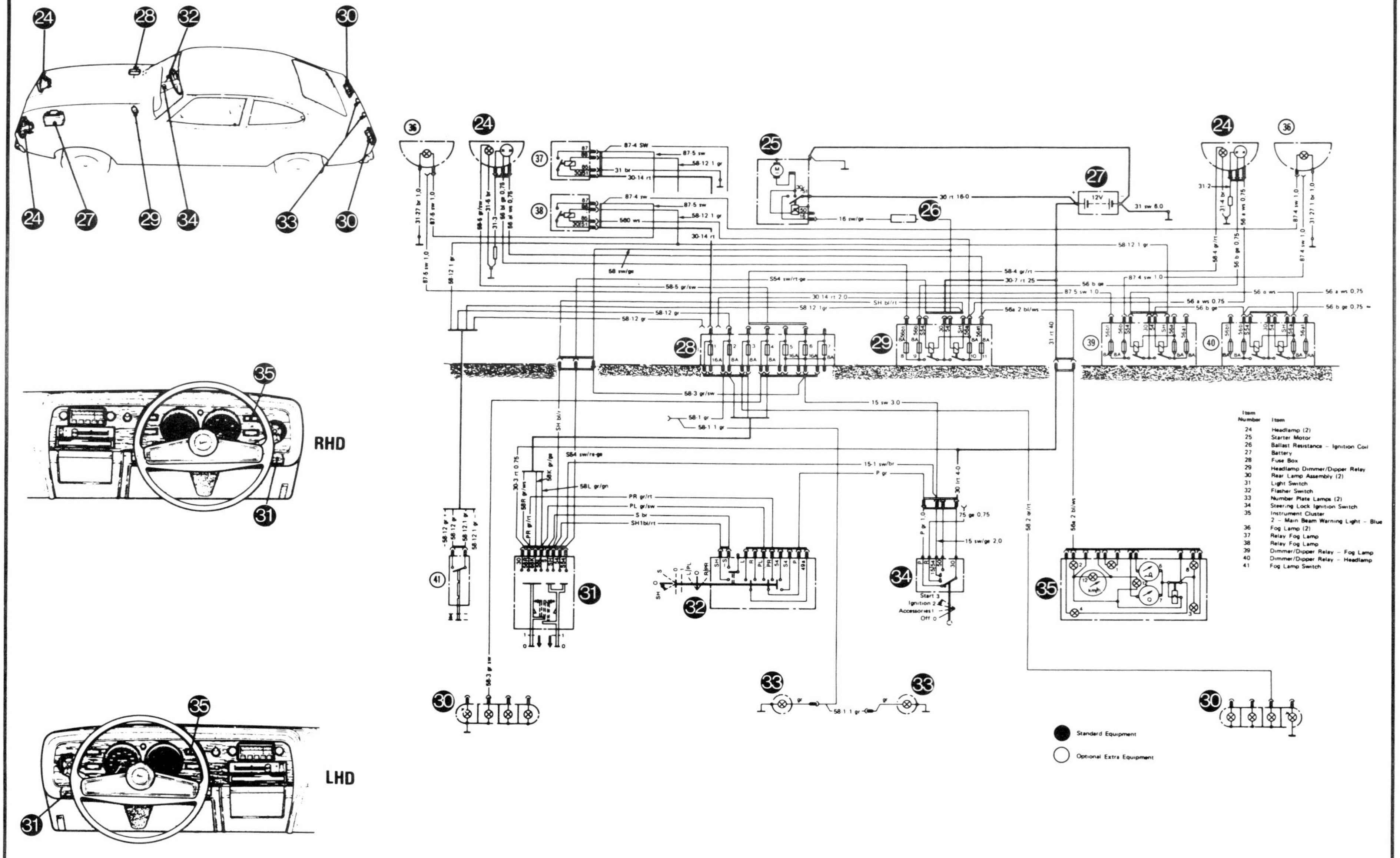

Item Number	Item
24	Headlamp (2)
25	Starter Motor
26	Ballast Resistance – Ignition Coil
27	Battery
28	Fuse Box
29	Headlamp Dimmer/Dipper Relay
30	Rear Lamp Assembly (2)
31	Light Switch
32	Flasher Switch
33	Number Plate Lamps (2)
34	Steering Lock Ignition Switch
35	Instrument Cluster 2 – Main Beam Warning Light – Blue
36	Fog Lamp (2)
37	Relay Fog Lamp
38	Relay Fog Lamp
39	Dimmer/Dipper Relay – Fog Lamp
40	Dimmer/Dipper Relay – Headlamp
41	Fog Lamp Switch

Fig. 10.40. Exterior light circuits (see page 132 for wiring codes)

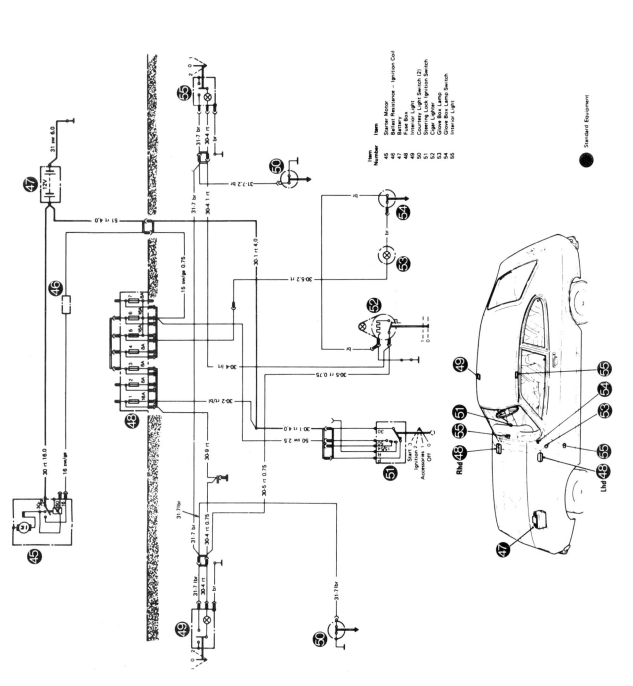

Fig. 10.41. Interior light circuits (see page 132 for wiring codes)

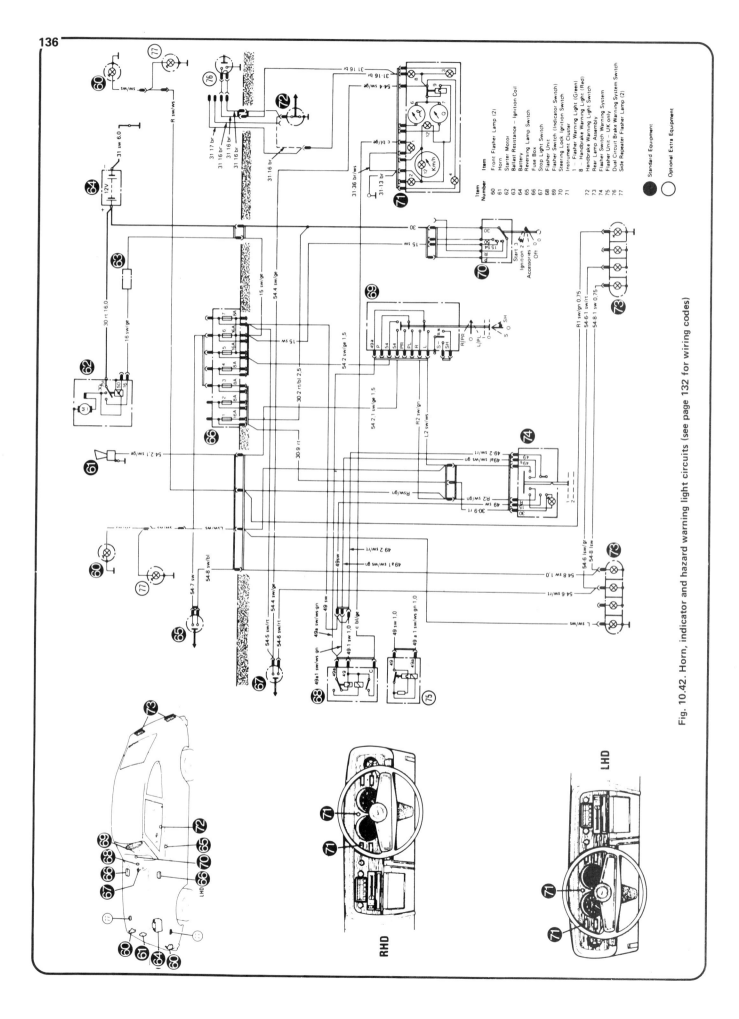

Fig. 10.42. Horn, indicator and hazard warning light circuits (see page 132 for wiring codes)

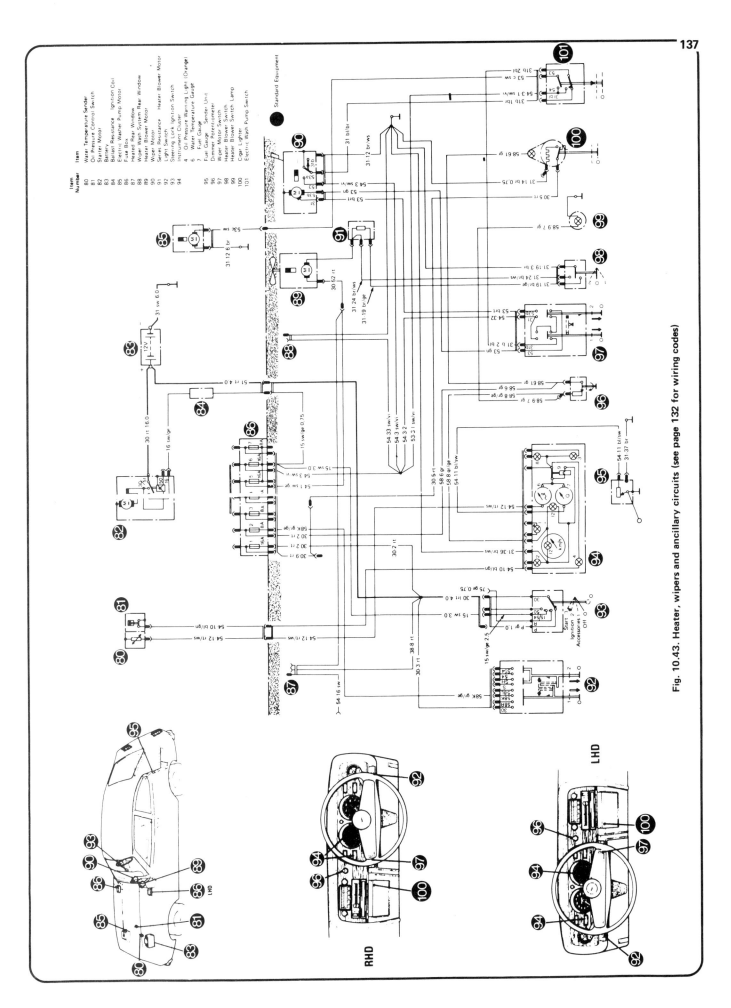

Fig. 10.43. Heater, wipers and ancillary circuits (see page 132 for wiring codes)

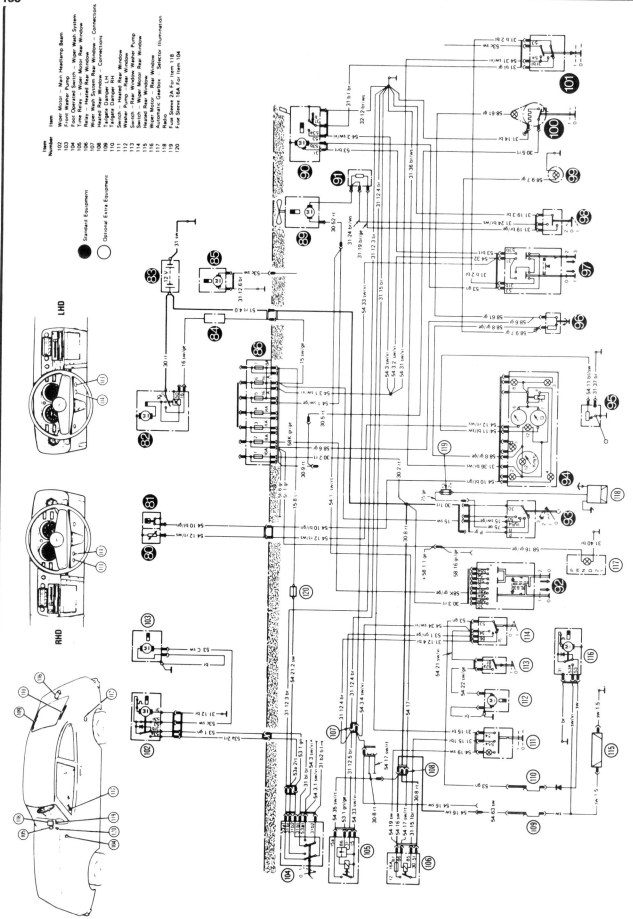

Fig. 10.44. Regular production option circuits additional to Fig. 10.43 (see page 132 for wiring codes)

Chapter 11 Suspension and steering

For modifications, and information applicable to later models, see Supplement at end of manual

Contents

Fault diagnosis - suspension and steering ... 17	Rear stabilizer bar - removal, renewal of bushes and refitting ... 9
Front coil spring - removal and refitting ... 5	Steering angles and front wheel alignment ... 15
Front hub bearings - maintenance and adjustment ... 2	Steering column - removal, dismantling, reassembly and refitting ... 14
Front hub - dismantling and bearing removal ... 3	Steering gear - adjustments ... 12
Front stabilizer bar - removal and refitting ... 6	Steering gear - dismantling, overhaul and reassembly ... 13
Front suspension strut - removal and refitting ... 4	Steering gear - removal and refitting ... 11
General description ... 1	Track control arm (suspension arm) - removal and refitting ... 7
Rear leaf spring - removal, renewal of bushes and refitting ... 10	Wheels and tyres ... 16
Rear shock absorber - removal and refitting ... 8	

Specifications

Front suspension
Type	Independent, MacPherson strut
Lateral control	Track control arms
Longitudinal control	Stabilizer bar
Shock absorbers	Hydraulic, telescopic, double-acting
Fluid type	SM6C-1003-A
Fluid capacity	325 ± 15 cc (0.18 Imp. pint, approx/0.22 US pint approx.)
Spring rating	The spring rating varies according to the vehicle and intended market. When replacements are required, consult a Ford dealer for further information

Rear suspension
Type	Semi-elliptic, leaf spring with rigid axle and stabilizer bar
Shock absorbers	Hydraulic, telescopic, double-acting
Spring rating	The spring rating varies according to the vehicle and intended market. When replacements are required, consult a Ford dealer for further information

Steering gear
Type	Rack and pinion
Rack travel (lock-to-lock)	5.08 in (129 mm)
Steering wheel turns (lock-to-lock)	3.36
Teeth on pinion (helical)	5
Lubricant type	SAE 90 EP gear oil
Lubricant capacity	0.15 litre (0.25 Imp. pint/0.3 US pint)
Steering gear adjustment	By shims
Pinion bearing shim thicknesses	0.005, 0.007, 0.010, 0.090 in (0.127, 0.178, 0.254, 2.286 mm)
Rack slipper bearing shim thicknesses	0.002, 0.005, 0.010, 0.015, 0.020 in (0.051, 0.127, 0.254, 0.381, 0.508 mm)

Front wheel alignment (unladen)
Castor angle	0° 33' to 1° 48'
Max. difference (side to side)	0° 45'
Camber angle	0° 15' to 1° 45'
Max. difference, side to side	1° 0'
Toe-in	0 to 0.28 in (0 to 7 mm)
Track (front)	53.3 in (1353 mm)

Wheels
L versions	Pressed steel

Tyres
Size	600 - 134PR or 185/70 HR13	
Pressures (600 - 134PR)	**Front**	**Rear**
Load up to 3 persons	21 lb/in^2 (1.5 kg/cm^2)	24 lb/in^2 (1.7 kg/cm^2)
Load in excess of 3 persons	24 lb/in^2 (1.7 kg/cm^2)	30 lb/in^2 (2.1 kg/cm^2)

Pressures (185/70 HR13)
  Load up to 3 persons ... 20 lb/in² (1.4 kg/cm²) 21 lb/in² (1.5 kg/cm²)
  Load in excess of 3 persons ... 28 lb/in² (2.0 kg/cm²) 28 lb/in² (2.0 kg/cm²)

Note 1: For sustained high speeds consult the tyre manufacturer or a Ford dealer.
Note 2: Where there is a tyre chart on the inside of the glove compartment door, refer to this for recommended tyre pressures and loads.

Torque wrench settings

	lb f ft	kg fm
Suspension unit upper mounting bolts	15 to 18	2 to 2.4
Spindle to top mount assembly*	29 to 33	4.1 to 4.6
Track control arm ball stud nut	30 to 35	4.2 to 4.9
Stabilizer bar attachment clamps**	21 to 24.3	2.9 to 3.4
Stabilizer bar to track control arm nut**	15 to 45	2.1 to 6.2
Track control arm inner bushing**	18 to 22	2.5 to 3.1
Front suspension crossmember to body sidemember	29 to 37	4.1 to 5.1
Shock absorber to rear axle	39 to 46	5.3 to 6.3
Shock absorber to floor assembly	20 to 24	2.7 to 3.3
Stabilizer bar to axle tube	29 to 37	4 to 5
Stabilizer bar to sidemember	26 to 30	3.5 to 4.1
Locknut on stabilizer bar end-piece	29 to 37	4 to 5
Spring U-bolts	18 to 27	2.5 to 3.6
Front of rear spring	26 to 30	3.5 to 4.1
Rear of rear spring	8 to 10	1.1 to 1.4
Steering arm to suspension unit	30 to 34	4.1 to 4.7
Steering gear to crossmember	15 to 18	2.1 to 2.5
Trackrod-end to steering arm	18 to 22	2.5 to 3.0
Coupling to pinion spline	12 to 15	1.7 to 2.1
Universal joint to steering shaft spline	17 to 22	2.3 to 3.0
Steering wheel to shaft	20 to 25	2.8 to 3.5
Steering column tube to pedal box	15 to 18	1.7 to 2.1

*These are to be tightened with the wheels in the 'straight-ahead' position and the weight of the car resting on its wheels. They are to be locked by punching the nut into the slot using a 0.10 in (3 mm) diameter ball ended punch
**These are to be tightened with the weight of the car resting on its wheels

1 General description

Each of the independent front suspension MacPherson strut units consists of a vertical strut enclosing a double acting damper surrounded by a coil spring.

The upper end of each strut is secured to the top of the wing valance under the bonnet by rubber mountings.

The wheel spindle carrying the brake assembly and wheel hub is forged integrally with the suspension unit foot.

The steering arms are connected to each unit which is in turn connected to trackrods and then to the rack and pinion steering gear.

The lower end of each suspension unit is located by a track control arm. A stabilising torsion bar is fitted between the outer ends of each track control arm and secured at the front to mountings on the body front member.

A rubber rebound stop is fitted inside each suspension unit thus preventing the spring becoming over-extended and jumping out of its mounting plates. Upward movement of the wheel is limited by the spring becoming fully compressed but this is damped by the addition of a rubber bump stop fitted around the suspension unit piston rod which comes into operation before the spring is fully compressed.

Whenever repairs have been carried out on a suspension unit it is essential to check the wheel alignment as the linkage could be altered which will affect the correct front wheel settings.

Every time the car goes over a bump, vertical movement of a front wheel pushes the damper body upwards against the combined resistance of the coil spring and the damper piston.

Hydraulic fluid in the damper is displaced and forced through the compression valve into the space between the inner and outer cylinder. On the downward movement of the suspension, the road spring forces the damper body downwards against the pressure of the hydraulic fluid which is forced back again through the rebound valve. In this way the natural oscillations of the spring are damped out and a comfortable ride is obtained.

On the front uprights it is worth noting that there is a shroud inside the coil spring which protects the machined surface of the piston rod from road dirt.

The steering gear is of the rack and pinion type and is located on the front crossmember by two 'U' shaped clamps. The pinion is connected to the steering column by a flexible coupling.

2 Front hub bearings - maintenance and adjustment

1 At the interval given in the Routine Maintenance Section at the beginning of the manual, clean and re-pack the front wheel bearings, then adjust them as described in the following paragraphs.
2 Apply the handbrake, jack up the front of the car and remove the roadwheels.
3 Disconnect the hydraulic brake at the union on the suspension unit and either plug the open ends of the pipes, or have a jar handy to catch the escaping fluid.
4 Bend back the locking tabs on the two bolts holding the brake caliper to the suspension unit, undo the bolts and remove the caliper.
5 By judicious tapping and levering remove the dust cap from the centre of the hub.
6 Remove the split pin from the nut retainer and undo the larger adjusting nut from the stub axle.
7 Withdraw the thrust washers and the outer tapered bearing.
8 Pull off the complete hub and disc assembly from the stub axle.
9 Carefully prise out the grease seal from the back of the hub assembly and remove the inner tapered bearing.
10 Carefully clean out the hub and wash the bearings with petrol making sure that no grease or oil is allowed to get onto the brake disc.
11 Working the grease well into the bearings fully pack the bearing cages and rollers with wheel bearing grease. **Note:** Leave the hub cavity half empty to allow for subsequent expansion of the grease.
12 To reassemble the hub assembly first fit the inner bearing and then gently tap the grease seal back into the hub. If the seal was at all damaged during removal a new one must be fitted.
13 Refit the hub and disc assembly on the stub axle and slide on the outer bearing and the thrust washer.
14 To adjust the bearings, first fully unscrew the hub nut. Then, using a torque wrench, tighten the nut to 27 lbf ft (3.7 kgf m) whilst rotating the hub.
15 Loosen the nut by two flats (120°).
16 Refit the nut retainer, a new split pin, and the dust cap.
17 Refit the caliper and connect up the hydraulic brake line.
18 Bleed the brakes as described in Chapter 9.
19 Refit the roadwheels and lower the car to the ground.

Chapter 11/Suspension and steering

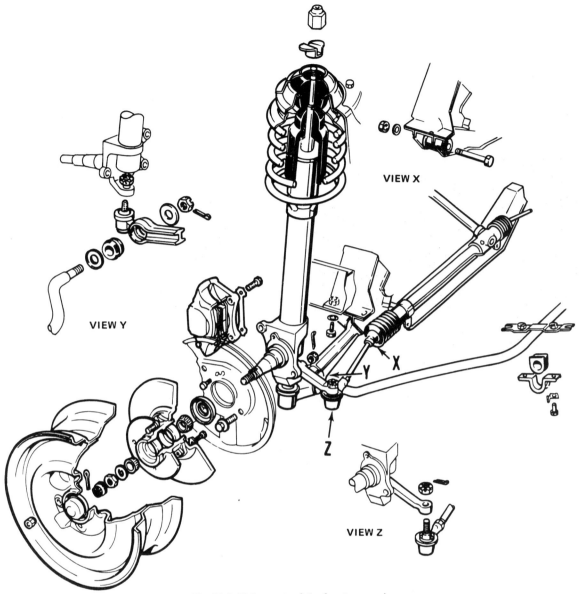

Fig. 11.1. Major parts of the front suspension

3 Front hub - dismantling and bearing renewal

1 Remove the hub/disc assembly, as described in Section 2.
2 Remove the oil seal and roller bearing races.
3 Using a brass drift or bearing puller, remove the bearing tracks from the ends of the hub.
4 If the disc is to be renewed because of scoring or distortion (see Chapter 9), bend down the lockplate tabs and unscrew and remove the bolts which connect the hub and disc.
5 Reassembly is a reversal of removal but if all front wheel bearings are being renewed, take care not to mix up the bearings and their tracks but keep them in their boxes until required as matched sets.
6 Use new locking plates under the disc to hub bolts and tighten all bolts to specifications.
7 Pack the bearings with grease and adjust them, as described in Section 2.

4 Front suspension strut - removal and refitting

1 It is difficult to work on the front suspension without one or two special tools, the most important of which is a set of adjustable spring clips which is Ford tool No. P.5045 (Fig. 11.2). This tool or similar clips or compressors are vital and any attempt to dismantle the units without them may result in personal injury.
2 Get someone to sit on the wing of the car and with the spring partially compressed in this way, securely fit the spring clips.
3 Jack up the car and remove the roadwheel, then disconnect the brake pipe at the bracket on the suspension leg and plug the pipes or have a jar handy to catch the escaping hydraulic fluid.
4 Disconnect the trackrod from the steering arm (see Section 11, paragraph 4), thus leaving the steering arm attached to the suspension unit.
5 Remove the outer end of the track control arm from the base of the suspension strut unit (for further information see Section 7).
6 Working under the bonnet, undo the three bolts holding the top end of the suspension strut to the side panel and lower the unit complete with the brake caliper away from the car.
7 Refitting is a direct reversal of the removal sequence but remember to use a new split pin on the steering arm to track rod nut and also on the track control arm to suspension unit nut.
8 The top suspension unit mounting bolts, the track control arm to suspension strut nut, and the steering arm to trackrod end nut must all be tightened to the specified torque.

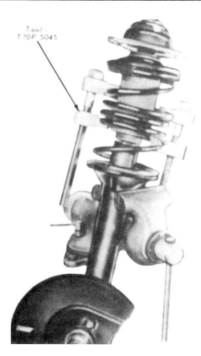

Fig. 11.2. Using a special tool to compress a front spring (Secs. 4 and 5)

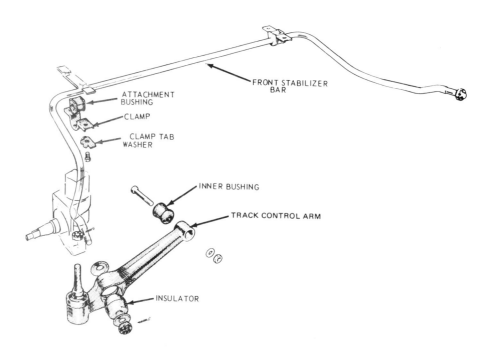

Fig. 11.3. The track control arm and stabilizing bar (Secs. 6 and 7)

5.14 Suspension unit cranked retainer - installed in position

5 Front coil spring - removal and refitting

1 Get someone to sit on the front wing of the car and with the spring partially compressed in this way securely fit spring clips or a road-spring compressor (see Fig. 11.2).
2 Jack up the front of the car, fit stands, and remove the roadwheel.
3 Working under the bonnet remove the piston nut and the cranked retainer.
4 Undo and remove the three bolts securing the top of the suspension unit to the side panel.
5 Push the piston rod downwards as far as it will go. It should now be possible to remove the top mounting assembly, the dished washer and the upper spring seat from the top of the spring.
6 The spring can now be lifted off its bottom seat and removed over the piston assembly.
7 If a new spring is being fitted check extremely carefully that it is of the same rating as the spring on the other side of the car. It is preferable to renew both springs at the same time. When ordering a replacement spring be sure to quote your car's serial number as well as the model, to ensure receiving the correct spring.
8 Before fitting a new spring it must be compressed with the adjustable restrainers and ensure that the clips are placed on the same number of coils, and in the same position as on the spring that has been removed.
9 Place the new spring over the piston and locate it on its bottom seat, then pull the piston up and fit the upper spring seat so that it locates correctly on the flats cut on the piston rod.
10 Fit the dished washer to the piston rod ensuring that the convex side faces upwards.
11 Now fit the top mounting assembly. With the steering in the straight-ahead position, fit the cranked retainer so that the ear on the retainer faces inwards and is at 90° to the centre-line of the car. Later models have retainers which incorporate two ears. Screw the piston rod nut on, having previously applied Loctite or a similar compound to the threads. Do not fully tighten the piston nut at this stage.
12 If necessary pull the top end of the unit upward until it is possible to locate correctly the top mounting bracket and fit the three retaining bolts. These bolts must be tightened down to the specified torque.
13 Remove the spring clips, fit the roadwheel and lower the car to the ground.
14 Finally slacken off the piston rod nut, get an assistant to hold the upper spring seat to prevent it turning and retighten the nut to the specified torque. Ensure that the cranked retainer faces inwards (ie. towards the engine) (photo).

6 Front stabilizer bar - removal and refitting

1 Jack up the front of the car, support the car on suitable stands and remove both front roadwheels.
2 Working under the car at the front, knock back the locking tabs on the four bolts securing the two front clamps that hold the stabilizer bar to the frame and then undo the four bolts and remove the clamps and rubber insulators (Fig. 11.3).
3 Remove the split pins from the castellated nuts retaining the stabilizer bar to the track control arms then undo the nuts and pull off the large washers, carefully noting the way in which they are fitted.
4 Pull the stabilizer bar forward out of the two track control arms and remove from the car.
5 With the stabilizer bar out of the car remove the sleeve and large washer from each end of the bar again noting the correct fitting

Chapter 11/Suspension and steering   143

positions.

6 Reassembly is a reversal of the above procedure, but new locking tabs must be used on the front clamp bolts and new split pins on the castellated nuts. The nuts on the clamps and the castellated nuts on each end of the stabilizer bar must not be fully tightened down until the car is resting on its wheels.

7 Once the car is on its wheels the castellated nuts on the ends of the stabilizer bar should be tightened down to the specified torque and the new split pins fitted. The four clamp bolts on the front mounting points must be tightened down to the specified torque and the locking tabs knocked up.

7 Track control arm (suspension arm) - removal and refitting

1 Jack up the front of the car, support it on suitable stands and remove the front wheel.
2 Working under the car remove the split pin and unscrew the castellated nut that secures the track control arm to the stabilizer bar.
3 Lift away the large dished washer noting which way round it is fitted.
4 Remove the self-locking nut and flat washer from the track control arm pivot bolt. Release the inner end of the track control arm.
5 Withdraw the split pin and unscrew the nut securing the track control arm ball joint to the base of the suspension unit. Separate the joint using a ball joint separator or wedges.
6 To refit the track control arm first assemble the track control arm ball stud to the base of the suspension unit.
7 Refit the nut and tighten to the specified torque. Secure with a new split pin.
8 Place the track control arm so that it correctly locates over the stabilizer bar and then secure the inner end.
9 Slide the pivot bolt into position from the front and secure with the flat washer and a new self-locking nut. The nut must be to the rear. Tighten the nut to the specified torque when the car is on the ground.
10 Fit the dished washer to the end of the stabilizer bar making sure it is the correct way round and secure with the castellated nut. This must be tightened when the car is on the ground, to the specified torque. Lock the castellated nut with a new split pin.

8 Rear shock absorber - removal and refitting

1 Remove the back seat after having removed the two screws from the floor assembly crossmember.
2 Remove the screws securing the seat belt to the top of the 'B' pillar.
3 Detach the 'B' pillar cover (2 screws).
4 Remove the top trim from the side window (4 screws).
5 Remove the two screws from the rocker panel at the rear end and pull off the door weatherstrip in the region of the side trim.
6 Take out the boot side trim (2 screws) and the carpet.
7 Remove the lining of the rear panel (5 screws) and of the side panel (10 screws) folding the rear seat forward for access.
8 Note the position of the steel and rubber washers at the wheel arch and axle mounting, then remove the shock absorber (Fig. 11.5).
9 Refitting is a direct reversal of the removal procedure, but ensure that the rubber and steel washers are correctly positioned (where these are showing signs of deterioration, replacement items should be used). Commence the refitting by first connecting the shock absorber at the axle and then extending it for fitting at the wheel arch end, and then reverse the above sequence to refit the panels.

9 Rear stabilizer bar - removal, renewal of bushes and refitting

1 Chock the front wheels to prevent the car moving, then jack up the rear of the car for access to the rear axle and stabilizer bar mountings.
2 Using a multi-grip wrench or similar tool to hold the stabilizer bar towards the axle tube, remove the two bolts at each stabilizer bar-to-axle tube bracket.
3 Disconnect the nut and bolt at each end of the stabilizer bar where it is attached to the floor assembly. Twist the stabilizer bar to remove it from the side.
4 To renew a stabilizer bar mounting bush, remove the locknut at one end and unscrew the end piece. Remove the nut and withdraw both rubber bushes from the stabilizer bar.
5 Dip the new rubber bushes in glycerine or brake fluid, ensure that the stabilizer bar surface is clean and not scored, then slide on the bushes and refit the end piece. When fitted, the end piece should be positioned as shown in Fig. 11.6 and the difference between the two sides must not be greater than 0.1 inch (2.5 mm).
6 If the bushes in the end pieces require renewal, it may be found more convenient to remove the end pieces from the stabilizer bar although this is not essential. The bushes can be pressed out using a suitable drift whilst the end piece is supported on a suitable diameter tube. Installation is straightforward, the new bushes being pressed in until the steel case on the outside of the bush is flush with the inside of the end piece. Note the position of the semi-circular recess in the bush as shown in Fig. 11.7.
7 When refitting the stabilizer bar it should be fitted at the floor end first with the washers and self-locking nuts loosely installed.
8 The bar is then fitted to the axle tube using a suitable tool to pull it towards the axle. The brackets, clamps and rubber insulators should now be fitted and the bolts tightened to the specified torque.

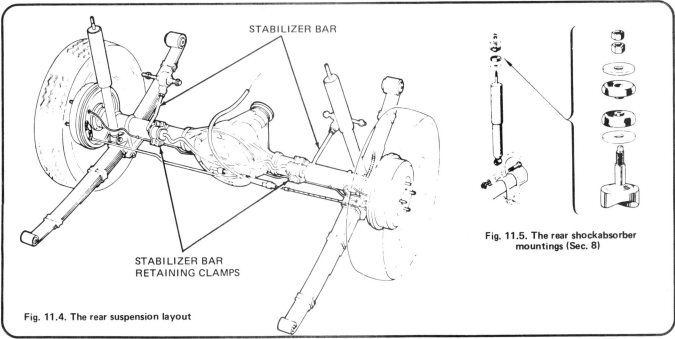

Fig. 11.5. The rear shockabsorber mountings (Sec. 8)

Fig. 11.4. The rear suspension layout

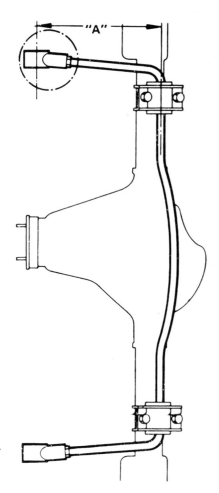

Fig. 11.6. The installation of the stabilizer bar (Sec. 9)

A = 262 mm ± 2.5 mm (10.314 in ± 0.1 in)

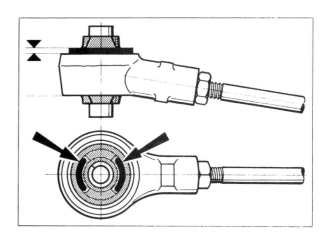

Fig. 11.7. The stabilizer bar rubber bush correctly positioned (Sec. 9)

9 Lower the vehicle to the ground and then load the vehicle so that the centre of the axle tube and the spring rear eye are on the same horizontal level (the weight required is approximately that of two adults). The nuts and bolts securing the stabilizer bar to the floor can now be torque tightened to the specified value.

10 Rear leaf spring - removal, renewal of bushes and refitting

1 Chock the front wheels to prevent the car moving, then jack up the rear of the car and support it on suitable stands. To make the springs more accessible remove the roadwheels.
2 Then place a trolley jack underneath the differential housing to support the rear axle assembly when the springs are removed. Do not raise the jack under the differential housing so that the springs are flattened, but raise it just enough to take the full weight of the axle with the springs fully extended.
3 Undo the rear shackle nuts and remove the combined shackle bolt and plate assemblies. Then remove the rubber bushes.
4 Undo the nut from the front mounting and take out the bolt running through the mounting.
5 Undo the nuts on the ends of the four 'U' bolts and remove the 'U' bolts together with the attachment plate and rubber spring insulators.

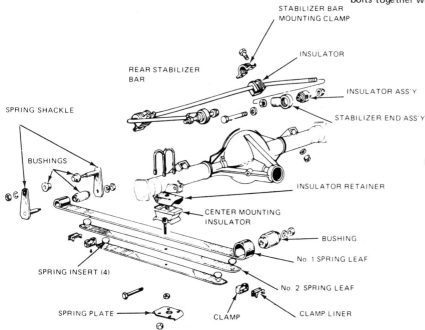

Fig. 11.8. Rear suspension - exploded view (Sec. 10)

Chapter 11/Suspension and steering

6 The rubber bushes can be pressed or driven out, and replacements fitted as described for the bushes in the stabilizer bar and end pieces in the previous Section. A little glycerine or brake fluid will allow the bushes to be pressed in more easily. Note that the front bushes are 7/16 inch (11 mm) diameter and the rear bushes are 5/16 inch (8 mm) diameter.

7 Refitting the spring is the reverse of the removal procedure. The nuts on the 'U' bolts, spring front mounting and rear shackles must be torqued down to the figures given in the Specifications at the beginning of this Chapter **after** the car has been lowered onto its wheels.

11 Steering gear - removal and refitting

1 Before starting this job, apply the handbrake and set the front wheels in the straight-ahead position. Then jack up the front of the car and place blocks under the wheels; lower the car slightly on the jack so that the trackrods are in a near horizontal position.

2 Remove the nut and bolt from the clamp at the front of the flexible coupling on the steering column. This clamp holds the coupling to the pinion splines (photo).

3 Working on the front crossmember, knock back the locking tabs on the two nuts on each rack housing 'U' clamp, undo the nut and remove the locking tabs and clamps.

4 Remove the split pins and castellated nuts from the ends of each trackrod where they join the steering arms. Separate the trackrods from the steering arms using a ball joint separator or wedges and lower the steering gear downwards out of the car.

5 Before refitting the steering gear make sure that the wheels have remained in the straight-ahead position. Also check the condition of the mounting rubbers round the housing and if they appear worn or damaged, renew them.

6 Check that the steering gear is also in the straight-ahead position. This can be done by ensuring that the distances between the ends of both trackrods and the steering gear housing on both sides are the same.

7 Place the steering gear in its location on the crossmember and at the same time mate up the splines on the pinion with the splines in the clamp on the steering column flexible coupling.

11.2 Steering column clamp nut and bolt (arrowed)

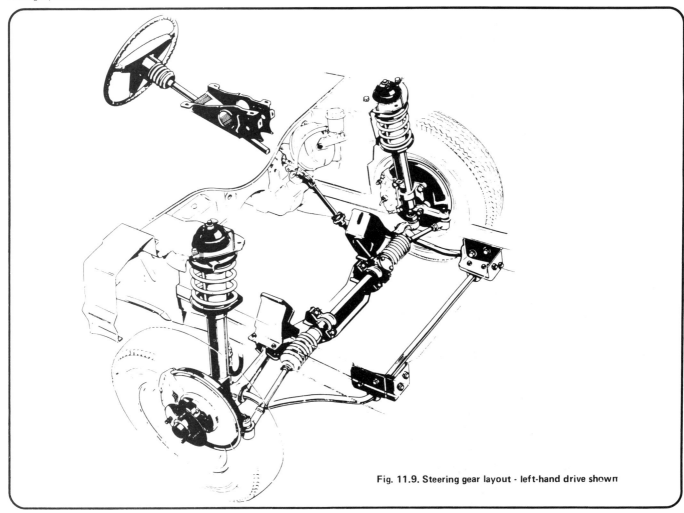

Fig. 11.9. Steering gear layout - left-hand drive shown

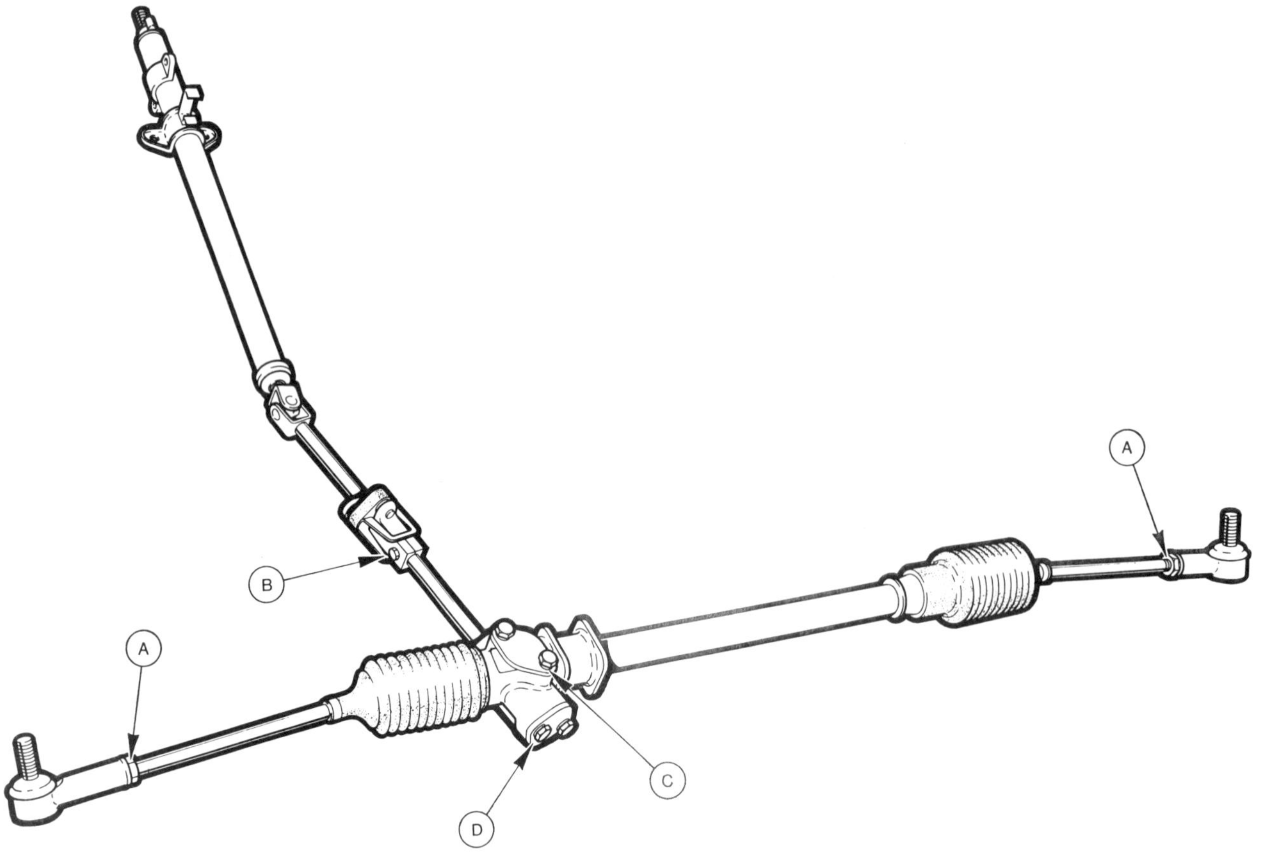

Fig. 11.10. Steering gear adjustment points (Sec. 11)

A Toe in adjustment
B Coupling shaft adjustment
C Rack slipper adjustment
D Pinion bearing adjustment

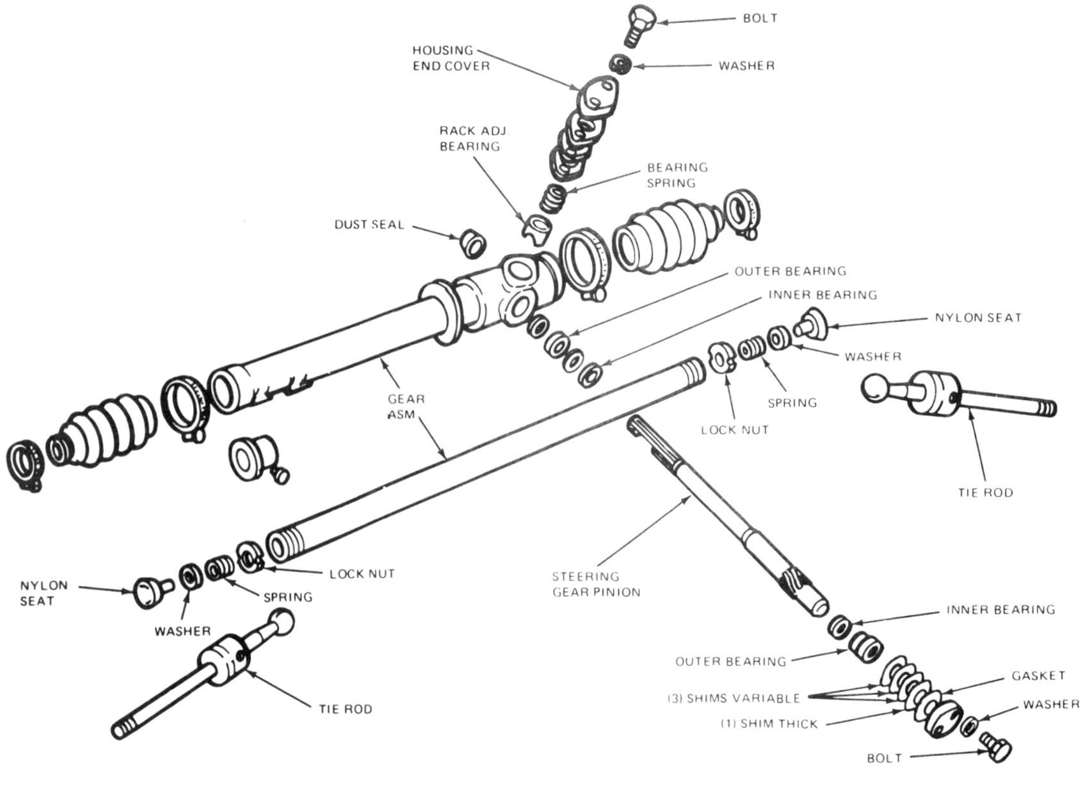

Fig. 11.11. Steering gear - component parts (Sec. 13)

8 Refit the two 'U' clamps using new locking tabs under the bolts, tighten down the bolts to the specified torque.
9 Refit the trackrod ends into the steering arms, refit the castellated nuts and tighten them to the specified torque. Use new split pins to retain the nuts.
10 Tighten the clamp bolt on the steering column flexible coupling to the specified torque, having first made sure that the pinion is correctly located in the splines.
11 Jack up the car, remove the blocks from under the wheels and lower the car to the ground. It is advisable at this stage to take the car to your local dealer and have the tracking checked (see Section 15).

12 Steering gear - adjustments

1 For the steering gear to function correctly, it is vital that the pinion bearing pre-load and the rack slipper are accurately adjusted. Ideally this will require the use of a dial gauge and mounting block, a surface table, a torque gauge and a splined adaptor. It is felt that most people will be able to suitably improvise using other equipment, but if this cannot be done and the equipment listed is not available, the job should be entrusted to your local vehicle main dealer.
2 To carry out these adjustments, remove the steering gear from the car as described in the previous Section. Mount the assembly in a soft jawed vice then remove the rack slipper cover plate, shim pack, gasket and spring.
3 Remove the pinion bearing cover plate, shim pack and gasket.

Pinion bearing preload

4 Place the shim pack and cover plate on the bearing, tighten the bolts then slacken them so that the cover plate touches the shim. The shim pack must comprise at least three shims, one of which must be 0.093 inch (2.35 mm), this being immediately against the cover plate.
5 Measure the cover plate-to-housing gap, and if outside the range 0.011 to 0.013 inch (0.28 to 0.33 mm) reduce the shim pack thickness (if the gap is too large) or increase it (if the gap is too small) until this gap is obtained. Remember that the 0.093 inch (2.35 mm) shim must remain immediately against the cover plate.
6 When the correct gap is obtained, remove the cover plate, install the gasket and refit the cover plate. Apply a sealer such as Loctite to the cover bolt threads, fit them and torque tighten to 6 to 8 lb f ft (0.83 to 1.1 kg fm).

Rack slipper adjustment

7 Having set the pinion bearing preload measure the height of the slipper above the main body of the rack as the rack is transversed from lock-to-lock by turning the pinion. Note the height reading obtained.
8 Prepare a shim pack which, including the thickness of the rack slipper bearing gasket, is 0.002 to 0.006 inch (0.05 to 0.15 mm) thicker than the dimension noted in paragraph 7.
9 Fit the spring, gasket, shim pack and cover plate to the rack housing (gasket nearest housing). Apply a sealer such as Loctite to the cover bolt threads, fit them and torque tighten to 6 to 8 lb f ft (0.83 to 1.1 kg fm).
10 Measure the torque required to turn the pinion throughout its range of travel. This should be 10 to 18 lb f in (11.5 to 20.7 kg cm); if outside this range, faulty components, lack of lubricant, etc., should be suspected.

13 Steering gear - dismantling, overhaul and reassembly

Note: The procedure given may be beyond the capabilities of many d-i-y motorists. Read through the Section before commencing any work and if not considered to be feasible, entrust the job to your local main dealer.

1 Remove and discard the wire retaining clips, remove the bellows and drain the lubricant.
2 Mount the steering gear in a soft-jawed vice and drill out the pins securing the trackrod housings to the locknuts. Centre-punch the pins before drilling then use a 4 mm (5/32 inch or No. 22) drill but do not drill too deeply.
3 It is now necessary to unscrew the housings from the ball joints so that the trackrods, housings, locknuts, ball seats, washers and springs can be removed. Ideally this requires the use of special tools which should be available from a main dealer but if improvised grips or

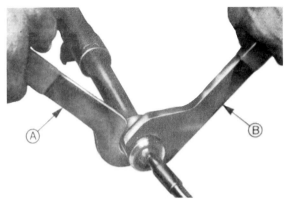

Fig. 11.12. Removing a track rod (Sec. 13)
A and B are special tools available for the purpose

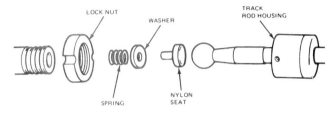

Fig. 11.13. Tie rod (track rod or connecting rod) ball joint exploded view (Sec. 13)

wrenches are used take care that no parts are damaged (if parts are damaged, replacement items must be obtained).
4 Remove the rack slipper cover plate, shim pack, gasket and slipper.
5 Remove the pinion bearing preload cover plate, shim pack, gasket and lower bearing.
6 Using a screwdriver or similar tool, prise out the pinion oil seal.
7 Clean all dirt and paint from the pinion shaft then push the pinion out of the housing.
8 Take out the pinion upper bearing and washer.
9 Clean and inspect all the parts for damage and wear. Examine the bush in the end of the rack tube furthest from the pinion; if worn it can be pressed out and a replacement fitted.
10 Commence reassembly by fitting the pinion upper bearing and washer into the housing.
11 Position the rack into the housing, and leave it in the central position.
12 Install the pinion, ensuring that after installation the flat is towards the right-hand side of the vehicle (irrespective of right or left-hand drive vehicles).
13 Fit the pinion lower bearing cover plate and adjust the preload as described in the previous Section.
14 Assemble the rack slipper, spring, gasket, shim pack and cover plate, adjusting as described in the previous Section.
15 Lubricate the ball seats, balls and housings with SAE 90 EP gear oil. Screw the locknuts onto the ends of the steering rack.
16 Assemble the springs, washers, ball seats, trackrod ends and housing. Tighten the housings to obtain a rotational torque of 5 lb f ft (0.7 kg fm) then lock them with the locknuts. Recheck the torque after tightening the locknut.
17 Drill new holes (even if the old holes are in alignment), 4 mm (5/32 inch or No. 22 drill) diameter x 9 mm (0.38 inch) deep along the break lines between the housing and the locknut, approximately opposite the spanner locating hole in the housing.
18 Fit new retaining pins and peen over the surrounding metal to retain them.
19 Lightly grease the inside of the bellows where they will contact the trackrods, install one bellow, ensuring that it locates in the trackrod groove; then fit a new retaining clip. Do not tighten the clip until the lock-in has been checked.
20 Add the specified quantity of steering gear oil, operating the rack over its range of travel to assist the lubricant in flowing. Do not overfill.
21 Fit the other bellow, but do not tighten the (new) clip yet.
22 Check the pinion turning torque, as described in paragraph 10 of the previous Section.

14 Steering column - removal, dismantling, reassembly and refitting

1 Disconnect the battery earth lead.
2 Remove the upper and lower steering coupling clamp bolts, and tap the coupling shaft down the pinion shaft to disconnect the coupling shaft from the column.
3 Carefully prise out the motif from the centre of the steering wheel and then unscrew the wheel retaining nut (photo).
4 Ensure that the roadwheels are in the straight-ahead position then pull off the steering wheel. Do not damage the collapsible can.
5 Remove the direction indicator actuator cam.
6 Remove the steering column shroud (2 screws at the bottom, then pull out at the top) and lower the dash panel trim.
7 Disconnect the direction indicator switch from the column (two bolts - see Fig. 11.14).
8 Disconnect the loom wiring from the ignition switch, noting the respective wiring positions.
9 Remove the two steering column retaining bolts (see Fig. 11.15). and pull the column assembly from the vehicle. Push the grommet out of the floor pan.
10 Drill off the steering column lockbolt heads, or tap them round with a pin punch, then use suitable grips to pull out the bolt shanks. Remove the steering lock (refer to Chapter 10, if necessary).
11 Remove the circlip snapring, washer and spring from the lower end of the column.
12 Tap the lower end of the shaft with a soft-faced hammer to remove the shaft and bearing from the top of the column.
13 Using the shaft as a drift, tap the lower bearing out of the column.
14 Inspect all the parts for wear and damage, renewing if necessary.
15 Commence reassembly by positioning the shaft in the column, then assemble the lower bearing (smaller diameter towards the column), spring, washer and circlip to the shaft. Push the assembly into the column to locate the bearing against the stops.
16 Press the upper bearing onto the column.
17 Secure the steering lock to the column and shear the bolts.
18 Use the steering lock to locate the shaft in the column, then fit the direction indicator actuating cam and steering wheel (check that the roadwheels are still in the 'straight-ahead' position).
19 Install the steering column grommet at the lower end.
20 Locate the column assembly and secure it with the two mounting bolts.
21 The remainder of refitting is the reverse of the removal procedure.

15 Steering angles and front wheel alignment

1 Accurate front wheel alignment is essential for good steering and tyre wear. Before considering the steering angle, check that the tyres are correctly inflated, that the front wheels are not buckled, that the hub bearings are not worn or incorrectly adjusted and that the steering linkage is in good order, without slackness or wear at the joints.
2 Wheel alignment consists of four factors:
Camber which is the angle at which the front wheels are set from the vertical when viewed from the front of the car. Positive camber is the amount (in degrees) that the wheels are tilted outwards at the top from the vertical.
Castor is the angle between the steering axis and a vertical line when viewed from each side of the car. Positive castor is when the steering axis is inclined rearwards.
Steering axis inclination is the angle when viewed from the front of the car, between the vertical and an imaginary line drawn between the

14.3 Removal of the steering wheel motif

Fig. 11.14. Retaining bolts for direction indicator switch (arrowed) (Sec. 14)

Fig. 11.15. Retaining bolts for steering column (arrowed) (Sec. 14)

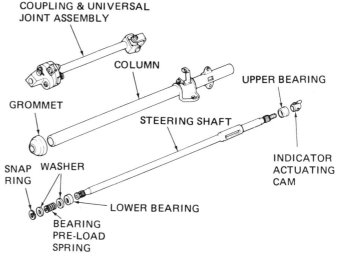

Fig. 11.16. The steering column assembly (Sec. 14)

Chapter 11/Suspension and steering

upper and lower suspension strut pivots.

Toe-in is the amount by which the distance between the **front** inside edges of the roadwheels (measured at hub height) is less than the distance measured between the **rear** inside edges.

3 The angles of camber, castor and steering axis are set in production and are not adjustable.

4 Front wheel alignment (toe-in) checks are best carried out with modern setting equipment but a reasonably accurate alternative is by means of the following procedure.

5 Place the car on level ground with the wheels in the 'straight-ahead' position.

6 Obtain or make a toe-in gauge. One may easily be made from a length of rod or tubing, cranked to clear the sump or bellhousing and having a setscrew and locknut at one end.

7 With the gauge, measure the distance between the two inner wheel rims at hub height at the front of the wheel.

8 Rotate the roadwheel through $180°$ (half a turn) by pushing or pulling the car and then measure the distance again at hub height between the inner wheel rims at the rear of the roadwheel. This measurement should either be the same as the one just taken or greater by not more than 0.28 inch (7 mm).

9 Where the toe-in is found to be incorrect slacken the locknuts on each trackrod, also the flexible bellows clips and rotate each trackrod by an equal amount until the correct toe-in is obtained. Tighten the trackrod end locknuts while the ball joints are held in the centre of their arcs of travel. It is imperative that the lengths of the trackrods are always equal otherwise the wheel angles on turns will be incorrect. If new components have been fitted, set the roadwheels in the 'straight-ahead' position and also centralise the steering wheel. Now adjust the lengths of the trackrods by turning them so that the trackrod end ball joint studs will drop easily into the eyes of the steering arms. Measure the distances between the centres of the ball joints and the grooves on the inner ends of the trackrods and adjust, if necessary so that they are equal. This is an initial setting only and precise adjustment must be carried out as described in earlier paragraphs of this Section.

16 Wheels and tyres

1 Check the tyre pressures weekly (when they are cold).

2 Frequently inspect the tyre walls and treads for damage and pick out any large stones which have become trapped in the tread pattern.

3 If the wheels and tyres have been balanced on the car then they should not be moved to a different axle position. If they have been balanced off the car then, in the interests of extending tread life, they can be moved between front and rear on the same side of the car and the spare incorporated in the rotational pattern.

4 Never mix tyres of different construction or very dissimilar tread patterns.

5 Always keep the roadwheels tightened to the specified torque and if the bolt holes become elongated or flattened, renew the wheel.

6 Occasionally, clean the inner faces of the roadwheels and if there is any sign of rust or corrosion, paint them with metal preservative paint. **Note:** Corrosion on aluminium alloy wheels (non-standard), may be evidence of a more serious problem which could lead to wheel failure. If corrosion is evident, consult your Ford dealer for advice.

7 Before removing a roadwheel which has been balanced on the car, always mark one wheel stud and bolt hole so that the roadwheel may be refitted in the same relative position to maintain the balance.

17 Fault diagnosis - suspension and steering

Before diagnosing faults from the following chart, check that any irregularities are not caused by:
1. *Binding brakes*
2. *Incorrect 'mix' of radial and crossply tyres*
3. *Incorrect tyre pressures*
4. *Misalignment of the bodyframe*

Symptom	Reason/s
Steering wheel can be moved considerably before any sign of movement of the roadwheels is apparent	Wear in the steering linkage, gear and column coupling.
Vehicle difficult to steer in a consistent straight line - wandering	As above. Wheel alignment incorrect (indicated by excessive or uneven tyre wear). Front wheel hub bearings loose or worn. Worn ball joints.
Steering stiff and heavy	Incorrect wheel alignment (indicated by excessive or uneven tyre wear). Excessive wear or seizure in one or more of the joints in the steering linkage or suspension. Excessive wear in the steering gear.
Wheel wobble and vibration	Roadwheels out of balance. Roadwheels buckled. Wheel alignment incorrect. Wear in the steering linkage, suspension ball joints or track control arm pivot. Broken front spring.
Excessive pitching and rolling on corners and during braking	Defective shock absorbers and/or broken spring.

Chapter 12 Bodywork and fittings

For modifications, and information applicable to later models, see Supplement at end of manual

Contents

Bonnet release cable - removal and refitting ... 24	Heater water valve (heavy duty heater) - removal and refitting ... 35
Bonnet - removal, refitting and adjustment ... 25	Instrument panel crash pad - removal and refitting ... 30
Bumpers - removal and refitting ... 8	Load space trim panel - removal and refitting ... 23
Centre console - removal and refitting ... 31	Maintenance - bodywork and underframe ... 2
Demister nozzles - removal and refitting ... 38	Maintenance - locks and hinges ... 7
Door lock assembly - removal and refitting ... 17	Maintenance - upholstery and carpets ... 3
Door rattles - tracing and rectification ... 12	Maintenance - vinyl roof covering ... 4
Door remote control handle - removal and refitting ... 13	Major body damage - repair ... 6
Door trim panel - removal and refitting ... 14	Minor body damage - repair ... 5
Door window frame - removal and refitting ... 19	Opening rear quarter glass assembly - removal and refitting ... 22
Door window glass - removal and refitting ... 18	Radiator grille - removal and refitting ... 9
Door window regulator assembly - removal and refitting ... 15	Rear quarter trim panel - removal and refitting ... 21
External door handle - removal and refitting ... 16	Seatbelts - general ... 32
Face level vents (vent registers) - removal and refitting ... 39	Tailgate assembly - removal and refitting ... 26
Fault diagnosis - heating system ... 40	Tailgate lock assembly - removal and refitting (including removal of the lock barrel) ... 27
Fuel filler flap - removal and refitting ... 29	Tailgate striker plate - removal and refitting ... 28
General description ... 1	Tailgate window glass - removal and refitting ... 11
Heater assembly - removal and refitting ... 36	Window frame mouldings and door weatherstrips - removal and refitting ... 20
Heater assembly (standard and heavy duty) - dismantling and reassembly ... 37	Windscreen - removal and refitting ... 10
Heater control - adjustments ... 33	
Heater controls - removal and refitting ... 34	

Specifications

Wheelbase ...	100.8 in (2559 mm)
Overall width ...	66.9 in (1698 mm)
Overall height ...	53 in (1357 mm)
Overall length	
With bumper overriders ...	170.9 in (4340 mm)
Without bumper overriders ...	168.8 in (4288 mm)

1 General description

The body is of a monocoque all-steel, welded construction with impact absorbing front and rear sections.

The car has two side doors and a full-length lifting tailgate for easy access to the rear compartment. The side doors are fitted with anti-burst locks and incorporate a key operated lock in each handle; window frames are adjustable for position. The tailgate hinges are bolted to the underside of the roof panel and to the tailgate itself. Gas-filled dampers support the tailgate in the open position; when closed it is fastened by a key-operated lock incorporating a release pushbutton.

An automatic bonnet locking mechanism operates when the bonnet is closed, a release lever being fitted at the edge of the instrument panel on the driver's side. The bonnet is hinged at the rear and is held in the open position by a support stay.

Toughened safety glass is fitted to all windows, the windscreen having an additional 'zone' toughened band in front of the driver. In the event of the windscreen shattering this zone crazes into large sections to give a greater degree of visibility as a safety feature. An optional glass/plastic/glass laminated windscreen is available at extra cost; this has the advantage of cracking only, to give an even greater degree of visibility in the event of accidental damage. The front door windows have a conventional winding mechanism. On certain variants, frameless opening rear quarter windows are fitted. These are hinged at the forward edge and are operated from an 'over-centre' type latch.

A padded facia crash panel is standard equipment together with deep pile wall-to-wall carpeting. Inertia reel seatbelts are fitted to all models.

To prevent damage under minor impacts, rubber faced bumpers are used.

All models are fitted with a heating and ventilating system which operates by ram air when the car is moving, or by a blower when stationary or for increased airflow. The heater is operated from a central control panel and airflow is directed to the windscreen or car interior according to the control lever settings.

2 Maintenance - bodywork and underframe

1 The condition of your car's bodywork is of considerable importance as it is upon this that the secondhand value of the car will mainly depend. It is very much more difficult to repair neglected bodywork than to renew mechanical assemblies. The hidden portions of the body, such as the wheel arches and the underframe and the engine compartment are equally important though obviously not requiring such frequent attention as the immediately visible paintwork.

2 Once a year or every 12,000 miles (19,000 km), it is sound advice to visit your local main agent and have the underside of the body steam cleaned. This will take about 1½ hours. All traces of dirt and oil will

Fig. 12.1. Body sheet metal panels (Sec. 1)

be removed and the underside can then be inspected carefully for rust, damaged hydraulic pipes, frayed electrical wiring and similar maladies.

3 At the same time the engine compartment should be cleaned in the same manner. If steam cleaning facilities are not available then brush a water soluble cleanser over the whole engine and engine compartment with a stiff paintbrush, working it well in where there is an accumulation of oil and dirt. Do not paint the ignition system but protect it with oily rags when the cleanser is washed off. As the cleanser is washed away it will take with it all traces of oil and dirt, leaving the engine looking clean and bright.

4 The wheel arches should be given particular attention as undersealing can easily come away here and stones and dirt thrown up from the roadwheels can soon cause the paint to chip and flake, and so allow rust to set in. If rust is found, clean down to the bare metal with wet or dry paper, paint on an anti-corrosive coating and renew the paintwork and undercoating.

5 The bodywork should be washed once a week or when dirty. Thoroughly wet the car to soften the dirt and then wash the car down with a soft sponge and plenty of clean water. If the surplus dirt is not washed off very gently, in time it will wear the paint down as surely as wet or dry paper. It is best to use a hose if this is available. Give the car a final wash down and then dry with a soft chamois leather to prevent the formation of spots.

6 Spots of tar and grease thrown up from the road can be removed with a rag dampened with petrol.

7 Once every six months, or every three months, if wished, give the bodywork and chromium trim a thoroughly good wax polish. If a chromium cleaner is used to remove rust or clean any of the car's plated parts remember that it also removes part of the chromium plating so use sparingly.

3 Maintenance - upholstery and carpets

1 Remove the carpets and thoroughly vacuum clean the interior of the car every three months or more frequently if necessary.

2 Beat out the carpets and vacuum clean them if they are very dirty. If the headlining or upholstery is soiled apply an upholstery cleaner with a damp sponge and wipe off with a clean dry cloth.

4 Maintenance - vinyl roof covering

Under no circumstances try to clean any external vinyl roof covering with detergents, caustic soaps or spirit cleaners. Plain soap and water is all that is required with a soft brush to clean dirt that may be ingrained. Wash the covering as frequently as the rest of the car.

5 Minor body damage - repair

See photo sequences on pages 158 and 159.

Repair of minor scratches in the car's bodywork

If the scratch is very superficial and does not penetrate to the metal of the bodywork, repair is very simple. Lightly rub the area of the scratch with a paintwork renovator or a very fine cutting paste, to remove loose paint from the scratch and to clear the surrounding bodywork of wax polish. Rinse the area with clean water.

Apply touch-up paint to the scratch using a thin paintbrush; continue to apply thin layers of paint until the surface of the paint in the scratch is level with the surrounding paintwork. Allow the new paint at least two weeks to harden; then blend it into the surrounding paintwork by rubbing the paintwork in the scratch area with a paintwork renovator or a very fine cutting paste. Finally apply wax polish.

Where the scratch has penetrated right through to the metal of the bodywork, causing the metal to rust, a different repair technique is required. Remove any loose rust from the bottom of the scratch with a penknife, then apply rust inhibiting paint to prevent the formation of rust in the future. Using a rubber or nylon applicator, fill the scratch with bodystopper paste. If required, this paste can be mixed with cellulose thinners to provide a very thin paste which is ideal for filling narrow scratches. Before the stopper paste in the scratch hardens, wrap a piece of smooth cotton rag around the top of a finger. Dip the finger in cellulose thinners and then quickly sweep it across the surface of the stopper paste in the scratch; this will ensure that the surface of the stopper paste is slightly hollowed. The scratch can now be painted over as described earlier in this Section.

Repair of dents in the car's bodywork

When deep denting of the car's bodywork has taken place, the first task is to pull the dent out, until the affected bodywork almost attains its original shape. There is little point in trying to restore the original shape completely, as the metal in the damaged area will have stretched on impact and cannot be reshaped fully to its original contour. It is better to bring the level of the dent up to the point which is just below the level of the surrounding bodywork. In cases where the dent is very shallow anyway, it is not worth trying to pull it out at all.

If the underside of the dent is accessible, it can be hammered out gently from behind, using a mallet with a wooden or plastic head. Whilst doing this, hold a suitable block of wood firmly against the impact from the hammer blows and thus prevent a large area of bodywork from being 'belled-out'.

Should the dent be in a section of the bodywork which has a double skin or some other factor making it inaccessible from behind, a different technique is called for. Drill several small holes through the metal inside the dent area - particularly in the deeper sections. Then screw long self-tapping screws into the holes just sufficiently for them to gain a good purchase in the metal. Now the dent can be pulled out by pulling on the protruding heads of the screws with a pair of pliers.

The next stage of the repair is the removal of the paint from the damaged area, and from an inch or so of the surrounding 'sound' bodywork. This is accomplished most easily by using a wire brush or abrasive pad on a power drill, although it can be done just as effectively by hand using sheets of abrasive paper. To complete the preparations for filling, score the surface of the bare metal with a screwdriver or the tang of a file, or alternatively, drill small holes in the affected area. This will provide a really good 'key' for the filler paste.

To complete the repair see the Section on filling and respraying.

Repair of rust holes or gashes in the car's bodywork

Remove all paint from the affected area and from an inch or so of the surrounding 'sound' bodywork, using an abrasive pad or a wire brush on a power drill. If these are not available a few sheets of abrasive paper will do the job just as effectively. With the paint removed you will be able to gauge the severity of the corrosion and therefore decide whether to replace the whole panel (if this is possible) or to repair the affected area. Replacement body panels are not as expensive as most people think and it is often quicker and more satisfactory to fit a new panel than to attempt to repair large areas of corrosion.

Remove all fittings from the affected area except those which will act as a guide to the original shape of the damaged bodywork (eg headlamp shells etc.). Then, using tin snips or a hacksaw blade, remove all loose metal and any other metal badly affected by corrosion. Hammer the edges of the hole inwards in order to create a slight depression for the filler paste.

Wire brush the affected area to remove the powdery rust from the surface of the remaining metal. Paint the affected area with rust inhibiting paint; if the back of the rusted area is accessible treat this also, this also.

Before filling can take place it will be necessary to block the hole in some way. This can be achieved by the use of one of the following materials: Zinc gauze or Aluminium tape.

Zinc gauze is probably the best material to use for a large hole. Cut a piece to the approximate size and shape of the hole to be filled, then position it in the hole so that its edges are below the level of the surrounding bodywork. It can be retained in position by several blobs of filler paste around its periphery.

Aluminium tape should be used for small or very narrow holes. Pull a piece off the roll and trim it to the approximate size and shape required, then pull off the backing paper (if used) and stick the tape over the hole; it can be overlapped if the thickness of one piece is insufficient. Burnish down the edges of the tape with the handle of a screwdriver or similar, to ensure that the tape is securely attached to the metal underneath.

Bodywork repairs - filling and re-spraying

Before using this Section, see the Sections on dent, deep scratch, rust hole, and gash repairs.

Many types of bodyfiller are available, but generally speaking those proprietary kits which contain a tin of filler paste and a tube of resin hardener are best for this type of repair. A wide, flexible plastic or nylon applicator will be found invaluable for imparting a smooth and well contoured finish to the surface of the filler.

Mix up a little filler on a clean piece of card or board — measure the hardener carefully (follow the maker's instructions on the pack) otherwise the filler will set too rapidly or too slowly.

Using the applicator, apply the filler paste to the prepared area; draw the applicator across the surface of the filler to achieve the correct contour and to level the filler surface. As soon as a contour that approximates to the correct one is achieved, stop working the paste - if you carry on too long the paste will become sticky and begin to 'pick-up' on the applicator. Continue to add thin layers of filler paste at twenty-minute intervals until the level of the filler is just 'proud' of the surrounding bodywork.

Once the filler has hardened, excess can be removed using a metal plane or file. From then on, progressively finer grades of abrasive paper should be used, starting with a 40 grade production paper and finishing with 400 grade 'wet-or-dry' paper. Always wrap the abrasive paper around a flat rubber, cork, or wooden block - otherwise the surface of the filler will not be completely flat. During the smoothing of the filler surface the 'wet-or-dry' paper should be periodically rinsed in water. This will ensure that a very smooth finish is imparted to the filler at the final stage.

At this stage the 'repair area' should be surrounded by a ring of bare metal, which in turn should be encircled by the finely 'feathered' edge of the good paintwork. Rinse the repair area with clean water, until all of the dust produced by the rubbing-down operation is gone.

Spray the whole repair area with a light coat of grey primer - this will show up any imperfections in the surface of the filler. Repair these imperfections with fresh filler paste or bodystopper, and once more smooth the surface with abrasive paper. If bodystopper is used, it can be mixed with cellulose thinners to form a really thin paste which is ideal for filling small holes. Repeat this spray and repair procedure until you are satisfied that the surface of the filler, and the feathered edge of the paintwork are perfect. Clean the repair area with clean water and allow to dry fully.

The repair area is now ready for spraying. Paint spraying must be carried out in a warm, dry, windless and dust free atmosphere. This condition can be created artificially if you have access to a large indoor working area, but if you are forced to work in the open, you will have to choose your day very carefully. If you are working indoors, dousing the floor in the work area with water will 'lay' the dust which would otherwise be in the atmosphere. If the repair area is confined to one body panel, mask off the surrounding panels; this will help to minimise the effects of a slight mis-match in paint colours. Bodywork fittings (eg. chrome strips, door handles etc.), will also need to be masked off. Use genuine masking tape and several thicknesses of newspaper for the masking operation.

Before commencing to spray, agitate the aerosol can thoroughly, then spray a test area (an old tin, or similar) until the technique is mastered. Cover the repair area with a thick coat of primer; the thickness should be built up using several thin layers of paint rather than one thick one. Using 400 grade 'wet-or-dry' paper, rub down the surface of the primer until it is really smooth. While doing this, the work area should be thoroughly doused with water, and the 'wet-or-dry' paper periodically rinsed in water. Allow to dry before spraying on more paint.

Spray on the top coat, again building up the thickness by using several thin layers of paint. Start spraying in the centre of the repair area and then, using a circular motion, work outwards until the whole repair area and about 2 inches of the surrounding original paintwork is covered. Remove all masking material 10 to 15 minutes after spraying on the final coat of paint.

Allow the new paint at least 2 weeks to harden fully; then, using a paintwork renovator or a very fine cutting paste, blend the edges of the new paint into the existing paintwork. Finally, apply wax polish.

Chapter 12/Bodywork and fittings

6 Major body damage - repair

1 Because the body is built on the monocoque principle and is integral with the underframe, major damage must be repaired by competent mechanics with the necessary welding and hydraulic straightening equipment.
2 If the damage has been serious it is vital that the body is checked for correct alignment as otherwise the handling of the car will suffer and many other faults such as excessive tyre wear and wear in the transmission and steering may occur.
3 There is a special body jig which most large body repair shops have and to ensure that all is correct it is important that this jig be used for all major repair work.

7 Maintenance - locks and hinges

Once every 6 months or 6,000 miles (10,000 km) the door, bonnet and tailgate hinges should be lubricated with a few drops of engine oil. Door striker plates can be given a thin smear of grease to reduce wear and ensure free movement.

8 Bumpers - removal and refitting

Front bumper
1 Initially disconnect the battery earth lead.
2 Remove the radiator cover which is retained by five screws.
3 Remove the four bumper retaining screws and lift away the bumper.
4 Refitting is the reverse of the removal procedure, but do not fully tighten the bolts until the bumper is correctly aligned.

Rear bumper
5 Open the tailgate, then remove the mat and the sub-floor.
6 Remove the jack and washer water reservoir (where applicable).
7 Remove two nuts, spring and flat washers at each end and lift away the bumper.
8 Refitting is the reverse of the removal procedure, but do not fully tighten the nuts until the bumper is correctly aligned.

Bumper trim strips
9 The bumper trim strips can be removed, and replacements fitted, by prising them out of, and pushing new ones into, the retaining grooves. The job is made a little easier if a soap and water solution is applied to the T-shaped retaining groove.

9 Radiator grille - removal and refitting

1 Initially disconnect the battery earth lead.
2 Remove the eight screws and washers and lift away the grille.
3 When refitting, ensure that the eight special nuts are correctly positioned on the front and lower crossmembers, then align and secure the bumper with the screws and washers.
4 Reconnect the battery earth lead.

10 Windscreen - removal and refitting

1 If you are unfortunate enough to have a windscreen shatter, or should you wish to renew your present windscreen, fitting a replacement is one of the few jobs which the average owner is advised to leave to a professional but for the owner who wishes to attempt the job himself the following instructions are given.
2 Cover the bonnet with a blanket or cloth to prevent accidental damage and remove the windscreen wiper blades and arms as detailed in Chapter 10.
3 Put on a pair of lightweight shoes and get into one of the front seats. With a piece of soft cloth between the soles of your shoes and the windscreen glass, place both feet in one top corner of the windscreen and push firmly. (See Fig. 12.4).

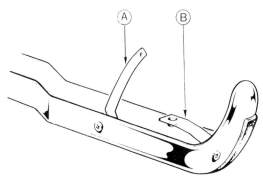

Fig. 12.2. Front bumper mounting (Sec. 8)

A Mounting bracket inner   B Mounting bracket outer

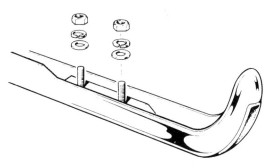

Fig. 12.3. Rear bumper mounting (Sec. 8)

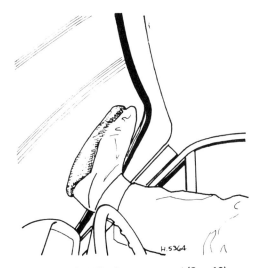

Fig. 12.4. Windscreen removal (Sec. 10)

4 When the weatherstrip has freed itself from the body flange in that area, repeat the process at frequent intervals along the top edge of the windscreen until, from outside the car, the glass and weatherstrip can be removed together. An assistant will be required to lift the screen clear on removal from the body flange.
5 If you are having to replace your windscreen due to a shattered screen, remove all traces of sealing compound and broken glass from the weatherstrip and body flange.
6 Gently prise out the clip which covers the joint of the chromium finisher strip and pull the finisher strip out of the weatherstrip. Then remove the weatherstrip from the glass or, if it is still on the car (as in

the case of a shattered screen) remove it from the body flange.

7 To fit a new windscreen start by fitting the weatherstrip around the new windscreen glass.

8 Apply a suitable sealer to the weatherstrip to body groove. In this groove then fit a fine but strong piece of cord right the way round the groove allowing an overlap of about 6 in (15 cm) at the joint.

9 From outside the car place the windscreen in its correct position making sure that the loose end of the cord is inside the car.

10 With an assistant pressing firmly on the outside of the windscreen get into the car and slowly pull out the cord thus drawing the weatherstrip over the body flange. (See Fig. 12.5).

11 Apply a further layer of sealer to the underside of the rubber to glass groove from outside the car.

12 Replace the chromium finisher strip into its groove in the weatherstrip and replace the clip which covers its joint.

13 Carefully clean off any surplus sealer from the windscreen glass before it has a chance to harden and then replace the windscreen wiper arms and blades.

11 Tailgate window glass - removal and refitting

1 Where applicable, remove the window glass wiper arm and blade, and carefully disconnect the heater element connections.

2 Carefully prise out the mylar insert from the rubber moulding.

3 If possible, obtain help from an assistant and carefully use a blunt bladed screwdriver to push the weatherstrip lip along the upper transverse section under the tailgate aperture flange. When approximately two thirds of the weatherstrip lip has been treated in this manner, pressure should be applied to the glass from inside the car. The glass and weatherstrip can then be removed from the outside.

4 Clean the lip of the window aperture, and the glass and weatherstrip if they are to be used again. Do not use solvents such as petrol or white spirit on the weatherstrip as this may cause deterioration of the rubber.

5 When refitting, initially fit the weatherstrip to the glass then insert a drawcord in the rubber-to-body groove so that the cord ends emerge at the bottom centre with approximately 6 in (15 cm) of overlap. During this operation it may help to retain the weatherstrip to the glass by using short lengths of masking tape.

6 Apply a suitable sealer to the body flange. Position the glass and weatherstrip assembly to the body aperture and push up until the weatherstrip groove engages the top transverse flange of the body aperture. Ensure that the ends of the draw cord are inside the car, then get the assistant to push the window firmly at the base whilst one end of the draw cord is pulled from the weatherstrip groove. Ensure that the cord is pulled at right-angles to the flange (ie. towards the centre of the glass) and that pressure is always being applied on the outside of the glass in the vicinity of the point where the draw cord is being pulled.

7 When the glass is in position, remove any masking tape which may have been used then seal the weatherstrip to the glass.

8 Lubricate the mylar insert with a rubber lubricant and refit it.

9 Refit the wiper arm and blade (where applicable), and reconnect the heater element connections.

12 Door rattles - tracing and rectification

1 The most common cause of door rattles is a misaligned, loose or worn striker plate. However, other causes may be:

    a) Loose door or window winder handles.
    b) Loose or misaligned door lock components.
    c) Loose or worn remote control mechanism.

2 It is quite possible for rattles to be the result of a combination of the above faults so a careful examination should be made to determine the exact cause.

3 If it is found necessary to adjust the striker plate, close the door to the first of the two locking positions. Visually check the relative attitude of the striker outside edge to the lock support plate edge. The edges 'A' and 'B' (Fig 12.6) should be parallel and can be checked by shining a torch through the door gap from above and below the striker.

4 Also check the amount by which the door stands proud of the

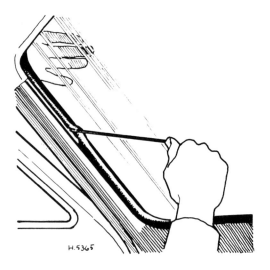

Fig. 12.5. Windscreen fitting (Sec. 10)

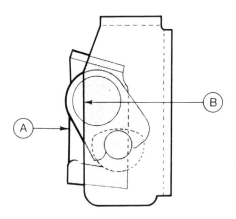

Fig. 12.6. Aligning the striker to lock support plate (Sec. 12)

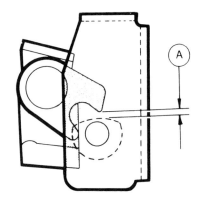

Fig. 12.7. Lock claw to striker clearance (Sec. 12)

adjacent panel. Adjust the striker plate as necessary to obtain a dimension of 0.24 in (6 mm).

5 With the lock in the open position check the lock claw striker clearance (dimension 'A' in Fig. 12.7). This should be 0.28 in (7 mm) and can be checked by placing a small ball of plasticine or similar on the striker post and checking its height after gently closing the door. The striker plate can be repositioned vertically to obtain this dimension but take care not to disturb any previous initial settings of the plate.

13 Door remote control handle - removal and refitting

1 Remove the door trim panel, as described in Section 14.
2 Push the remote control handle assembly towards the front of the car and pull it out of the opening in the door inner panel.
3 Lift the protective cap off the rear and twist the assembly to disengage it from the operating rod (photos).
4 Remove the protective cap from the operating rod (photo).
5 Refitting is the reverse of the removal procedure.

14 Door trim panel - removal and refitting

1 Carefully lift up and remove the window winder handle insert strip.
2 Remove the winder handle retaining screw and pull off the handle and escutcheon.
3 Remove the two armrest retaining screws, turn the armrest through 90° and pull out the top fixing.
4 Carefully prise out the remote control bezel and unscrew the private lock button (photo).
5 Taking care that no damage to the panel or paintwork occurs, carefully prise the trim panel from the door panel.
6 When refitting, press in the panel so that it is secured by its clips.
7 Refit the lock button, then position the bezel on the door remote control housing, push the trim pad clear of the housing and push the bezel rearwards to secure.
8 Position the spacer over the armrest stud. Position the armrest to the door and push the stud to secure it. Secure the armrest with the two screws.
9 Assemble the escutcheon over the winder shaft and install the winder so that when the window is closed the winder is in the lower vertical position. Secure the winder with the screw and refit the insert strip.

15 Door window regulator assembly - removal and refitting

1 Remove the door trim panel, as described in the previous Section.
2 Peel off the plastic sheet.
3 Temporarily refit the winder handle and lower the window. Remove the four gear plate fixing screws and the three pivot plate screws.

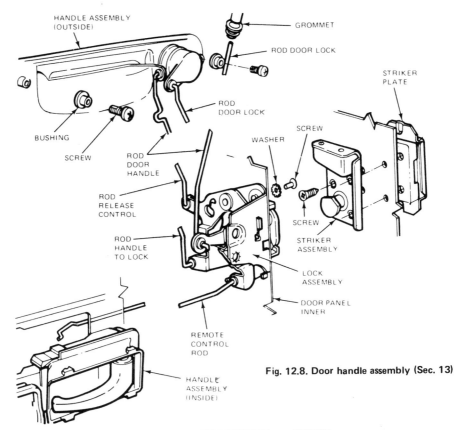

Fig. 12.8. Door handle assembly (Sec. 13)

13.3a Lift up the protective cap from the rear of the remote control handle ...

13.3b ... and twist the assembly to disengage it from the operating rod

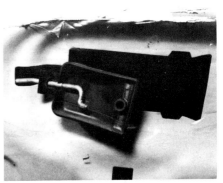

13.4 The remote control handle protective cap

Chapter 12/Bodywork and fittings

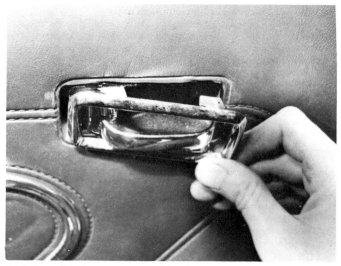

14.4 Removing the remote control bezel

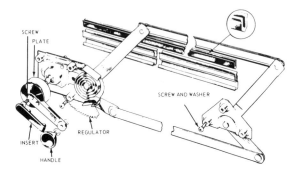

Fig. 12.9. Door window regulator assembly (Sec. 15)

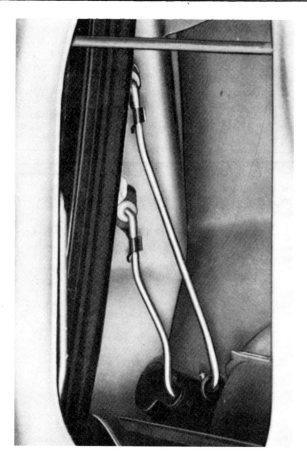

Fig. 12.10. Door lock to handle connecting links (Sec. 16)

4 Draw the regulator assembly towards the rear of the door to disengage it from the runner at the base of the window.
5 Push the window glass up and use adhesive tape on each side of the glass and over the window frame to retain it. If it is to be left for any length of time, additionally use a wooden support.
6 Withdraw the regulator from the door.
7 Installation is the reverse of the removal procedure, alignment being obtained by adjusting the pivot plate as necessary.

16 Exterior door handle - removal and refitting

1 Remove the door trim panel, as previously described.
2 Pull back the plastic sheet behind the exterior handle then disconnect the two connecting links from the door lock to the exterior handle.
3 Remove the two handle retaining bolts and withdraw the handle.
4 Installation is the reverse of the removal procedure but do not forget to install the bushes for the link rods. A little petroleum jelly on the rod ends will assist with their installation.

17 Door lock assembly - removal and refitting

1 Remove the door trim panel as previously described, and remove the plastic sheeting.
2 Remove the remote control handle and two window frame bolts.
3 Using a screwdriver, prise the clips from the exterior handle rod and detach the rods from the lock.
4 Remove the crosshead screws securing the lock to the shell and the plastic clips securing the remote control rod to the inner panel.
5 Remove the lock from the door through the lower rear access aperture.
6 When refitting, insert the remote control rod through the door aperture ensuring that the rod lies against the door inner panel. Locate

Fig. 12.11. The black (A), and white (B), door lock bushes (Sec. 17)

the lock on the door shell, pushing the frame towards the outer panel to enable the lock to be correctly positioned on the rear shell.
7 Secure the lock with the three screws, and the remote control rod to the inner panel with the two plastic clips.
8 Replace the exterior handle rods in their respective lock locations. Position the black bush 'A' and white bushes 'B' as shown in Fig. 12.11.
9 The remainder of refitting is the reverse of the removal procedure.

18 Door window glass - removal and refitting

1 Remove the door trim panel, as previously described, then peel the plastic sheeting away from the door panel apertures.

2 Remove the door belt moulding/weatherstrip assembly (see Fig. 12.12).
3 Wind up the window glass then remove the pivot plate screws. Remove the four regulator gear plate securing screws, then disengage the studs and rollers of the regulator arms from the door glass channel and carefully lift out the glass (Fig. 12.13). Allow the regulator to fall away, pivoting on the regulator handle shaft.
4 When refitting, initially insert a small block of wood in the bottom of the door assembly. Locate the glass in the door panel so that it is resting on the wooden block.
5 Locate the studs and rollers of the regulator arm into the door glass channel then temporarily install the winder handle and turn it to align the gear plate with the panel fixings. Secure the plate to the inner panel.
6 Loosely assemble the pivot plate then wind up the glass and align it in the frame. Tighten the pivot plate screws.
7 The remainder of refitting is the reverse of the removal procedure.

19 Door window frame - removal and refitting

1 Remove the door trim panel as previously described then peel the plastic sheeting away from the lower door panel apertures.
2 Remove the door bolt moulding/weatherstrip assembly (see Fig. 12.12).
3 Lower the window glass, then peel back the lower front cover of the plastic sheeting and remove the reflector (where applicable) to gain access to the front and rear lower fixing bolts.
4 Remove the five bolts and frame seals to free the frame from the shell. Push the glass out of the frame at the rear of the door so that the frame lies between the glass and the outer panel. Repeat for the front of the door.
5 Pull the rear of the frame from the shell whilst guiding the front of the frame rearwards past the first door bolt moulding retaining clip to enable the frame to be lifted clear.
6 When refitting, insert the front of the frame so that the vertical leg lies to the rear of the first moulding clip.
7 Spring the rear of the frame into the shell so that the frame lies between the glass and the outer panel whilst springing the frame front vertical leg past the moulding clip so that this also lies between the glass and the inner panel.
8 Spring the frame around the glass and secure it with the five bolts.
9 Pull the weatherstrip from the door aperture flange, then shut the door and adjust the frame to obtain a gap between the frame and flange (in and out) of 0.4 to 0.56 in (10 to 14 mm) and between the frame and the 'A' pillar (fore and aft) of 0.32 to 0.48 in (8 to 12 mm). Tighten the bolts.
10 The remainder of refitting is the reverse of the removal procedure.

20 Window frame mouldings and door weatherstrips - removal and refitting

1 Where applicable wind the window down to its fullest extent. Carefully prise the weatherstrip out of the groove in the door outer bright metal finish moulding.
2 When refitting, correctly position the weatherstrip over its groove. With the thumbs, carefully press the strip fully into the groove.
3 Wind the window up (where applicable) and check that the weatherstrip is correctly fitted.

21 Rear quarter trim panel - removal and refitting

1 Remove the screws retaining the window quarter trim and lift away the trim.
2 Remove the 'B' pillar vertical trim and the seatbelt screw (where applicable) (photo).

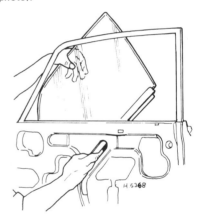

Fig. 12.13. Door window glass removal (Sec. 18)

Fig. 12.12. Removing the door moulding (Secs. 18 and 19)

Fig. 12.14. Window frame retaining bolts at A, B and as indicated by arrows (Sec. 19)

21.2 Removing the 'B' pillar vertical trim

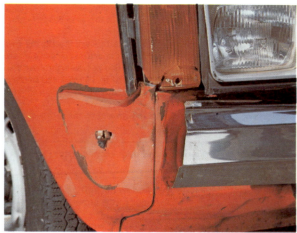

This sequence of photographs deals with the repair of the dent and paintwork damage shown in this photo. The procedure will be similar for the repair of a hole. It should be noted that the procedures given here are simplified — more explicit instructions will be found in the text

In the case of a dent the first job — after removing surrounding trim — is to hammer out the dent where access is possible. This will minimise filling. Here, the large dent having been hammered out, the damaged area is being made slightly concave

Now all paint must be removed from the damaged area, by rubbing with coarse abrasive paper. Alternatively, a wire brush or abrasive pad can be used in a power drill. Where the repair area meets good paintwork, the edge of the paintwork should be 'feathered', using a finer grade of abrasive paper

In the case of a hole caused by rusting, all damaged sheet-metal should be cut away before proceeding to this stage. Here, the damaged area is being treated with rust remover and inhibitor before being filled

Mix the body filler according to its manufacturer's instructions. In the case of corrosion damage, it will be necessary to block off any large holes before filling — this can be done with aluminium or plastic mesh, or aluminium tape. Make sure the area is absolutely clean before ...

... applying the filler. Filler should be applied with a flexible applicator, as shown, for best results; the wooden spatula being used for confined areas. Apply thin layers of filler at 20-minute intervals, until the surface of the filler is slightly proud of the surrounding bodywork

Initial shaping can be done with a Surform plane or Dreadnought file. Then, using progressively finer grades of wet-and-dry paper, wrapped around a sanding block, and copious amounts of clean water, rub down the filler until really smooth and flat. Again, feather the edges of adjoining paintwork

The whole repair area can now be sprayed or brush-painted with primer. If spraying, ensure adjoining areas are protected from over-spray. Note that at least one inch of the surrounding sound paintwork should be coated with primer. Primer has a 'thick' consistency, so will find small imperfections

Again, using plenty of water, rub down the primer with a fine grade wet-and-dry paper (400 grade is probably best) until it is really smooth and well blended into the surrounding paintwork. Any remaining imperfections can now be filled by carefully applied knifing stopper paste

When the stopper has hardened, rub down the repair area again before applying the final coat of primer. Before rubbing down this last coat of primer, ensure the repair area is blemish-free – use more stopper if necessary. To ensure that the surface of the primer is really smooth use some finishing compound

The top coat can now be applied. When working out of doors, pick a dry, warm and wind-free day. Ensure surrounding areas are protected from over-spray. Agitate the aerosol thoroughly, then spray the centre of the repair area, working outwards with a circular motion. Apply the paint as several thin coats

After a period of about two weeks, which the paint needs to harden fully, the surface of the repaired area can be 'cut' with a mild cutting compound prior to wax polishing. When carrying out bodywork repairs, remember that the quality of the finished job is proportional to the time and effort expended

3 Remove the two screws (A in Fig. 12.15) and remove the rear seat cushion. Where applicable, feed the seatbelt and buckle assemblies through the opening in the cushion.
4 Remove the three trim panel screws and the step plate. Also, where applicable, remove the luggage compartment hook. Carefully prise away the trim panel.
5 Installation is the reverse of the removal procedure, but on completion tighten the seatbelt bolt to a torque of 15 to 20 lb f ft (2.1 to 2.9 kg fm).

22 Opening rear quarter glass assembly - removal and refitting

1 Remove the trim covers from the 'B' panel and the quarter window surround (trim).
2 Remove the two toggle retaining screws and remove the toggle from the rear 'C' pillar.
3 Remove the window frame weatherstrip then drive out the toggle-to-catch retaining pin, remove the toggle.
4 Refitting is the reverse of the removal procedure, but lubricate the 'B' pillar hinge pivots with a soap solution prior to fitting the glass assembly. Adjust the toggle or weatherstrip flange to achieve 0.32 to 0.39 in (8 to 10 mm) gap between the glass and the weatherstrip flange.

23 Load space trim panel - removal and refitting

1 Remove the rear quarter trim panel, as previously described.
2 Pull the seat forward and remove the ten securing screws. It may also be necessary to detach the back trim panel (five screws).
3 Remove the panel after removing the interior light connection and the seatback lock knob.
4 Refitting is the reverse of the removal procedure. Ensure that the sound deadening material is correctly positioned and that the trim panel does not foul the seat release hinge mechanism.

24 Bonnet (hood) release cable - removal and refitting

1 In the event of the release cable breaking it is possible to remove the radiator grille to operate the lock spring by hand. Grille removal is dealt with in Section 9, but since it is not possible to open the bonnet it will be found a little difficult (though not impossible) to gain access to the upper retaining screws.
2 To remove the release cable in normal circumstances, remove the radiator upper cowl panel.
3 From inside the car remove both the clevis pins and the spring, and disconnect the release cable from the control lever.
4 Slacken the cable adjuster clamp and release the cable from the hood lock spring.
5 Remove the cable retaining clips then pull the cable through the dashpanel to remove it.
6 Refitting of the cable is essentially the reverse of the removal procedure, adjusting, as necessary, to remove any cable slack. For further information on this refer to paragraph 6, of the following Section.

25 Bonnet (hood) - removal, refitting and adjustment

1 Open the bonnet to its fullest extent. Using a suitable implement scribe a line around the hinges.
2 Remove the two bolts and washers on each side securing the bonnet to its hinges. With assistance it can now be lifted off.
3 Refitting is a reversal of the removal procedure. However, before fully tightening the securing bolts, ensure that the hinges are aligned with the scribed marks. This will ensure correct alignment.
4 If it is found that the bonnet requires adjustment, this can be effected in the vertical plane by slackening the catch post locknut and screwing the catch post in or out. Fore-and-aft adjustment can be effected by slackening the hinge bolts.
5 Adjustable bump rubbers are also provided, and these should be positioned as necessary to stop vibration but at the same time must allow the bonnet to be closed easily.
6 Adjustment of the bonnet locking spring can be made by slackening

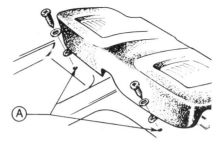

Fig. 12.15. Seat cushion securing screws (Sec. 21)

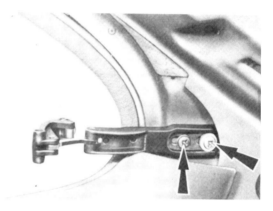

Fig. 12.16. Quarter window fixing screws (Sec. 22)

Fig. 12.17. Load space trim panel fixing points (Sec. 23)

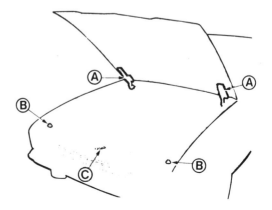

Fig. 12.18. Bonnet (hood) adjustment points (Sec. 25)

A Hinge   B Bumper rubber   C Striker

Chapter 12/Bodywork and fittings

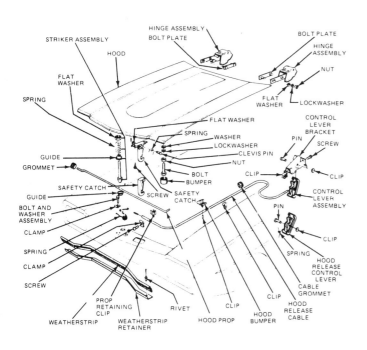

Fig. 12.19. Bonnet assembly - component parts (Sec. 25)

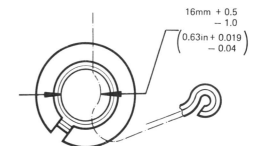

Fig. 12.20. Hood lock spring setting dimension (Sec. 25)

Fig. 12.21. The damper (strut) in-line connectors (Sec. 26)

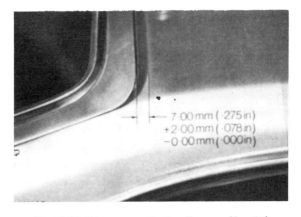

Fig. 12.22. Tailgate to roof edge alignment (Sec. 26)

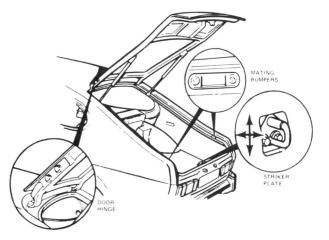

Fig. 12.23. Tailgate adjustment points (Secs. 26 and 28)

the cable clamp on the upper crossmember and sliding the outer cable through the clip as necessary. When correctly positioned, the hood lock spring/cable setting dimension should be as shown in Fig. 12.20. Tighten the clamp screw when the adjustment is satisfactory.

26 Tailgate assembly - removal and refitting

1 Open the tailgate and detach the inline connectors from the damper(s) strut(s).
2 Detach each damper by removing the securing bolt at each end.
3 With help from an assistant, support the tailgate and remove it by removing the hinge bolts.
4 When installing, align the tailgate so that the edges are flush with the rear of the roof and the 'C' pillar sides, and the gap between the tailgate and the roof edge is 0.275 to 0.353 in (7 to 9 mm) - see Fig. 12.22.
5 Align the lower edges of the tailgate so that it is flush with the rear corners of the body with the striker plate in the upper central position.
6 Pads can be used beneath the 'C' pillar bumpers for alignment of the tailgate sides with the 'C' pillar slope. Note that the thick end of the bumper faces towards the front of the vehicle.
7 On completion, refit the dampers. Assemble the spacer to the screw, then locate the screw and spacer through the pushrod end of the damper. Secure the damper to the 'C' pillar bracket, with the terminals facing rearwards.
8 Align the damper to the tailgate bracket and secure it with the remaining screw and spacer.
9 Reconnect the electrical connections to the dampers.

27 Tailgate lock assembly - removal and refitting (including removal of the lock barrel)

1 Open the tailgate and remove the latch (three bolts and washers).
2 Remove the lock cylinder retaining nut, then turn the lock spider to detach it from the lock cylinder and outer panel.
3 Installation is the reverse of the removal procedure.
4 If it is found necessary to remove the lock barrel, this can be done after it has been removed by removing the circlip (snap-ring) from the barrel housing. The spring, barrel and spider can then be detached and a replacement barrel fitted by reversing this procedure.

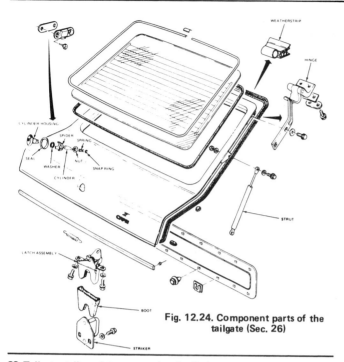

Fig. 12.24. Component parts of the tailgate (Sec. 26)

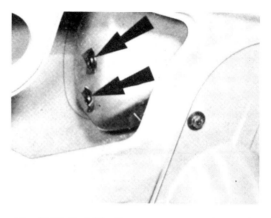

Fig. 12.25. Fuel filler flap retaining screws (Sec. 29)

28 Tailgate striker plate - removal and refitting

1 Open the tailgate then carefully scribe a mark around the striker to facilitate refitting.
2 Remove the single bolt and washer and take off the striker plate.
3 Refitting is the reverse of the removal procedure, following which adjustment can be made if found necessary to obtain satisfactory opening and closing of the tailgate.

29 Fuel filler flap - removal and refitting

1 Remove the loadspace trim panel, as previously described.
2 Remove the two screws indicated in Fig. 12.25 and lift away the filler flap.
3 Installation is the reverse of the removal procedure.

30 Instrument panel crash pad - removal and refitting

1 Disconnect the battery earth lead.
2 Remove the steering column shroud retaining screws. Remove the lower half and release the upper half retaining lug from its spring clip by pulling sharply upwards.
3 Remove the instrument cluster, as described in Chapter 10.
4 Detach the flexible pipes from the dashpanel vents and defrosters.
5 Remove the left and right-hand 'A' pillar trims (2 screws each) - also the grab handle if fitted.
6 Remove the instrument panel pad retaining screws and lift the pad away. If appropriate, remove the dash panel vents and transfer them to the new instrument panel.
7 Installation is essentially the reverse of the removal procedure, but connect the flexible pipes to the vents and defrosters before the crash pad is fitted.

31 Centre console - removal and refitting

1 Lift the carpet from around the front of the console then push the clock and bezel out of the housing. Disconnect the clock leads.
2 Remove the two screws at the rear end and two more from the clock end to release the console.
3 Remove the gear lever knob or T-handle.
4 Lift off the console. As applicable, remove the clock mounting plate screws and/or gearshift lever boot.
5 Installation is the reverse of the removal procedure.

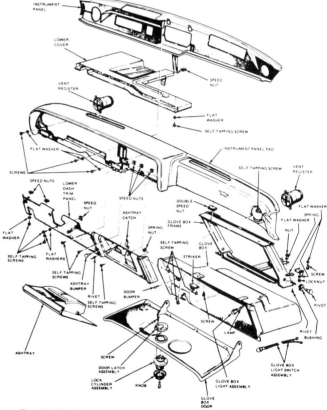

Fig. 12.26. Instrument panel crash pad and glove box (Sec. 30)

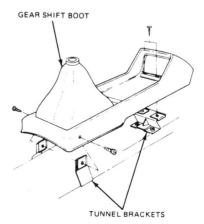

Fig. 12.27. Centre console (Sec. 31)

Chapter 12/Bodywork and fittings

32 Seatbelts - general

1 All models are fitted with inertia reel seatbelts for the front seats. Rear seatbelts of a similar type are also available.
2 Fig. 12.28 shows the floor mounted front seatbelt stalk. Removal of the basic type fixing is straightforward.
3 To remove the front seatbelt inertia reel, remove the rear quarter trim panel, as described in Section 21, then remove the three nuts from the anchor plate and lower the assembly through the aperture, whilst feeding the belt through the slot in the inner panel (photos).
4 If the seatbelt fixings are removed, they should be torque tightened to the following values on installation. (**Note**: This does not include the inertia reel anchor plate):

Front stalks	26 to 31 lb f ft (3.6 to 4.3 kg fm)
Other fixings	15 to 20 lb f ft (2.1 to 2.9 kg fm)

33 Heater controls - adjustment

1 Initially disconnect the battery earth lead, then remove the glove compartment by unscrewing 7 screws at the top and 2 nuts at the bottom. Also disconnect the glove compartment lighting leads.
2 Move the heater controls to a point 0.08 in (2 mm) from the end position, then remove both outer cable clips (see Figs. 12.31, 12.32 or 12.33 as appropriate).
3 *Standard heater:* Check that the distributor and regulator flap levers are at the end of their travel and clamp the outer cables in this position (Fig. 12.31).
4 *Heavy duty heater:* Check that the distributor flap lever (Fig. 12.32) and water control lever (Fig. 12.33) are at the end of their travel and clamp the outer cables in this position.
5 On completion, reconnect the glove compartment lighting leads, then refit the glove compartment and reconnect the battery earth lead.

32.3a Anchor plate stud nuts

32.3b Feeding the belt through the slot in the inner panel

32.3c Lowering the inertia reel through the slot in the inner panel

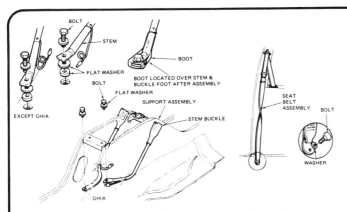

Fig. 12.28. Front seatbelt anchorage points (Sec. 32)

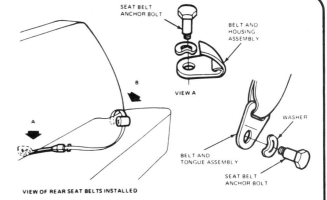

Fig. 12.29. Rear seatbelt anchorage points (Sec. 32)

Fig. 12.30. Removing the glove compartment (Sec. 33)

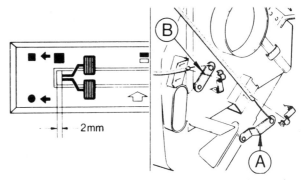

Fig. 12.31. Standard heater, control cable adjustment (Sec. 33)

A *Distributor flap control lever*   B *Regulator flap control lever*

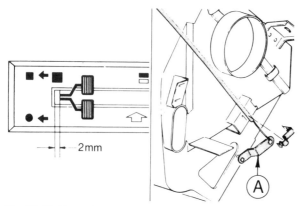

Fig. 12.32. Heavy duty heater, control cable adjustment (Sec. 33)

A Distributor flap control lever

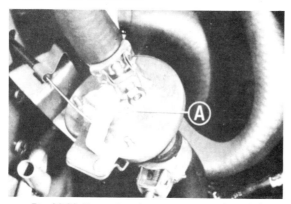

Fig. 12.33. Heavy duty heater water valve (Sec. 33)

A Valve control lever

34 Heater controls - removal and refitting

1 Initially disconnect the battery earth lead.
2 Remove the steering column shroud (2 screws at the bottom, then pull out at the top). Lower the steering column (leaving the two bolts in position), sufficiently to allow the instrument cluster trim to be removed.
3 Disconnect the switch leads and remove the instrument cluster trim complete with cowl trim (11 screws). Remove the instrument cluster bezel (3 screws).
4 Using a large pair of pliers, break the heater control knobs and remove the controls (4 screws). Do not disconnect the control cables at the heater.
5 Remove the heater control panel (2 screws). Remove the blower switch and lighting leads.
6 Disconnect the cables from the heater controls.
7 Refitting is the reverse of the removal procedure, during which it will be necessary to adjust the cables, as described in the previous Section. Also it will be necessary to obtain replacement heater control knobs.

35 Heater water valve (heavy duty heater) - removal and refitting

1 Drain the engine coolant and disconnect the lower hose from the radiator (refer to Chapter 2, if necessary).
2 Disconnect the three water hoses from the water valve (Fig. 12.34).
3 Remove the outer cable from the clip on the water valve bracket then remove the assembly from the bulkhead (2 screws).
4 Twist the water valve to disconnect the cable from the operating lever.
5 Installation is the reverse of the removal procedure, during which adjustment should be made, as described in Section 33 for the heavy duty heater. On completion, refit the radiator hose and fill the cooling system, as described in Chapter 2.

Fig. 12.34. The water valve hoses (arrowed) (Sec. 35)

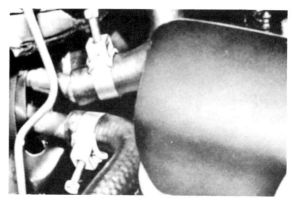

Fig. 12.35. Heat exchanger hoses (Sec. 36)

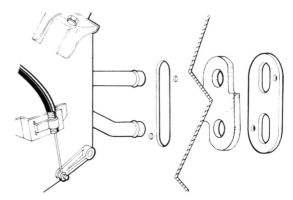

Fig. 12.36. Heat exchanger gasket and cover panel (Sec. 36)

36 Heater assembly - removal and refitting

1 Initially disconnect the battery earth lead.
2 Drain the coolant, referring to Chapter 2, if necessary.
3 Disconnect the water hoses from the heater heat exchanger. If practicable, blow through the heat exchanger with compressed air to remove any coolant remaining; alternatively place cloths and/or newspapers beneath to absorb any spillage.
4 Remove the cover panel together with the heat exchanger-to-water connection gasket from the bulkhead (2 screws) (Fig. 12.36).
5 Slacken the gearlever locknut then remove the gearlever. The locknut requires a special peg spanner available from Ford, but it is not difficult to fabricate a tool which will do the job.
6 Remove the parcel tray (4 screws); where there is a centre console this must be removed also (refer to Section 31).
7 Remove the steering column shroud (2 screws at the bottom, then pull out at the top). Lower the steering column (leaving the two bolts in position), sufficiently to allow the instrument cluster to be removed.
8 Remove the lower dash panel complete with cover panel (9 screws). Remove the ashtray and cigarette lighter, and disconnect the switches.

Chapter 12/Bodywork and fittings

9 Remove the glove compartment by unscrewing the 7 screws at the top and 2 nuts at the bottom. Also disconnect the glove compartment lighting leads.
10 Disconnect the demister nozzle hoses together with their connections (1 screw each).
11 Disconnect the facia vent hoses from the heater. There are 2 on the standard heater and 4 on the heavy duty heater.
12 Remove the lower dash panel support stay (1 screw).
13 Disconnect the control cables from the heater, and the heater blower leads.
14 Remove the demister nozzles (refer to Section 38, if necessary).
15 Remove the windscreen wiper motor bracket from its mounting.
16 Remove the 4 heater securing screws then pull the heater far enough rearward for the water connection pipes to clear the bulkhead. Tilt the top of the heater upward and forward, and withdraw it sideways; remove the foam gasket also.
17 Refitting is the reverse of the removal procedure, during which it will be necessary to adjust the heater controls as described in Section 33. Do not forget to tighten the gearlever locknut. On completion fill the cooling system, as described in Chapter 2.

37 Heater assembly (standard and heavy duty) - dismantling and reassembly

1 *Standard heater:* Remove clips 'A' and 'B' (Fig. 12.39) and remove

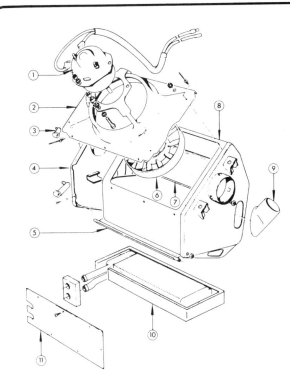

Fig. 12.37. Standard heater - component parts (Sec. 37)

1 Motor assembly
2 Cover and bracket assembly
3 Control valve operating lever
4 Right-hand housing cover
5 Distributor flap
6 Fan
7 Control valve
8 Housing
9 Demister hose connection
10 Heat exchanger seal
11 Plenum chamber cover

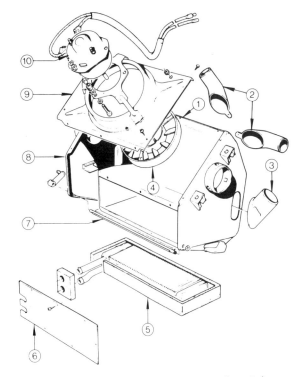

Fig. 12.38. Heavy duty heater - component parts (Sec. 37)

1 Housing
2 Hot air supply to facia connection
3 Demister hose connection
4 Fan
5 Heat exchanger seal
6 Plenum chamber cover
7 Distributor flap
8 Right-hand housing cover
9 Cover and bracket assembly
10 Motor assembly

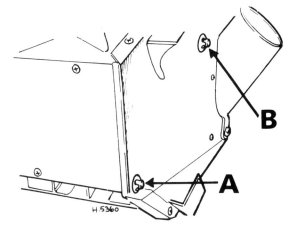

Fig. 12.39. Distributor flap (A) and regulator flap (B) clips (Sec. 37)

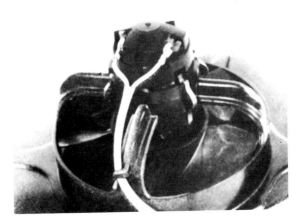

Fig. 12.40. The installed position of the blower motor (Sec. 37)

the heater housing side cover complete with flaps (15 screws).
2 *Heavy duty heater:* Remove clip 'A' (Fig. 12.39) and remove the heater housing side cover complete with flaps (15 screws).
3 Remove the heat-exchanger and foam seal.
4 Prise off the circlip and remove the fan from the blower motor shaft.
5 Detach the blower motor from the support (3 nuts and bolts).
6 Reassembly is the reverse of the dismantling procedure.

38 Demister nozzles - removal and refitting

Passenger's side
1 Remove the glove compartment by unscrewing the 7 screws at the top and 2 nuts at the bottom. Also disconnect the glove compartment lighting leads.
2 Withdraw the hose from the demister nozzle and remove the nozzle (1 screw).
3 Installation is the reverse of the removal procedure.

Driver's side
4 Initially proceed as described in paragraphs 1, 2 and 3 of Section 34. Additionally remove the ashtray and cigar lighter.
5 Remove the instrument cluster (4 screws), disconnecting the speedometer drive cable and electrical connections. If there is any doubt about the position of any of the electrical connections, make a note of them first of all.
6 Withdraw the demister nozzle hose and remove the demister nozzle (1 screw), turning it upward and outward so that the inlet side of the nozzle can come out first from the instrument cluster opening.
7 Refitting is the reverse of the removal procedure.

39 Face level vents (vent registers) - removal and refitting

Passenger's side
1 Remove the glove compartment by unscrewing 7 screws at the top and 2 nuts at the bottom. Also disconnect the glove compartment lighting leads.
2 Withdraw the hose(s) from the vent.
3 Remove the vent by unscrewing the 2 nuts which are accessible from the rear of the panel.
4 Installation is the reverse of the removal procedure.

Driver's side
5 Initially proceed as described in paragraphs 1, 2 and 3 of Section 34. Additionally remove the 9 screws at the top of the instrument panel trim.
6 Withdraw the hose(s) from the vent.

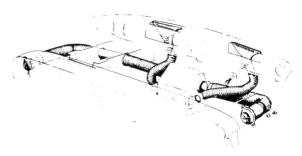

Fig. 12.41. Standard heater ducting (Secs. 39 and 40)

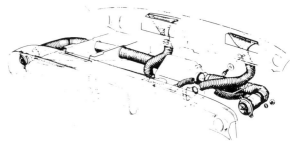

Fig. 12.42. Heavy duty heater ducting (Secs. 39 and 40)

7 Remove the vent by unscrewing the 2 nuts which are accessible from the rear of the panel.
8 Installation is the reverse of the removal procedure.

40 Fault diagnosis - heating system

Symptom	Reason
Insufficient heat	Faulty engine coolant reservoir cap.
	Faulty cooling system thermostat.
	Kink in heater hose.
	Faulty control lever or cable.
	Heat exchanger blocked.
	Blower fuse blown.
	Low engine coolant level.
Inadequate defrosting or general heat circulation	Incorrect setting of deflector doors.
	Disconnected ducts.
	Carpet obstructing airflow outlet.

Chapter 13 Supplement:
Revisions and information on later models

Contents

Introduction ... 1	Braking system ... 6
Specifications ... 2	Rear brake modification — description
Carburation ... 3	Rear drum brake shoes (modified) — inspection and renewal
Variable venturi (VV) carburettor — description	Rear drum brake wheel cylinder — removal and refitting
Variable venturi (VV) carburettor — adjustment	Rear drum brake wheel cylinder — servicing
Variable venturi (VV) carburettor — air filter element renewal	Handbrake — adjustment
Variable venturi (VV) carburettor — removal and refitting	Brake pedal free height — checking
Variable venturi (VV) carburettor — dismantling and reassembly	Fluid level warning indicator — description
Variable venturi (VV) carburettor — automatic choke adjustment	Teves (ATE) master cylinder — description
Variable venturi (VV) carburettor automatic choke — removal and refitting	Electrical system ... 7
Variable venturi (VV) carburettor — float level check	Electrical relays — description
Variable venturi (VV) carburettor — air valve adjustment	Headlamps/sidelights — bulb renewal
Clutch ... 4	Front direction indicator lights — bulb renewal
Release bearing — modification	Suspension and steering ... 8
Gearbox ... 5	Steering wheel — description
Speedometer drivegear — modification	Bodywork and fittings ... 9
Gear lever — modification	Window regulator knob — renewal

1 Introduction

With the exception of the modified braking system, this Supplement includes information which mainly applies to Series 3 models. The Series 3 was introduced in March 1978 and incorporated a four-headlamp system, a front spoiler and larger rear light clusters.

Mechanically the most significant innovation was the introduction in November 1979 of the variable venturi (VV) carburettor which gave better performance and improved economy.

In order to use the Supplement to the best advantage, it is suggested that reference is made to it before the main Chapters of the manual; this will ensure that any relevant information can be understood and incorporated into the procedures given in Chapters 1 to 12. Time and cost will therefore be saved and the particular job will be completed correctly.

2 Specifications

The specifications listed here are revised or supplementary to the main specifications given at the beginning of each Chapter

Carburation
Carburettor type (November 1979 on) ...	Ford variable venturi (VV)
Idle speed ...	800 ± 25 rpm
Idle CO % ...	1.75 ± 0.5
Operating fuel level ...	1.4 to 1.6 in (35 to 41 mm) from top of casting
Diaphragm lever setting dimension ...	0.47 in (12 mm)
Choke gauging twist drill diameter ...	0.135 in (3.4 mm)
Choke fast idle/pull-down twist drill diameter ...	0.177 in (4.5 mm)
Air valve maximum opening ...	0.9 in (23 mm)

Ignition system
Spark plugs
Type (engine code J3) ...	AGR 12

Distributor
Static advance (engines with VV carburettor) ...	10° BTDC

Braking system
Brake pedal free height ...	6.81 ± 0.40 in (173 ± 10 mm)

Electrical system
Fuses
Main fusebox on engine compartment bulkhead

Fuse and rating	Circuits protected
1 - 16 amp	Interior light(s), hazard flasher, cigarette lighters, clock, horn, headlamp washers
2 - 8 amp	Number plate lights, glove compartment light, instrument panel illumination
3 - 8 amp	RH tail and sidelights
4 - 8 amp	LH tail and sidelights
5 - 16 amp	Heater blower motor
6 - 16 amp	Reversing lights, wiper motors, screen wash systems
7 - 8 amp	Instrument cluster, stoplights
8 - 16 amp	LH dipped beam
9 - 16 amp	RH dipped beam
10 - 16 amp	RH main beam
11 - 16 amp	LH main beam

Additional fuses located below headlamp relay

Suspension and steering
Shock absorbers (front and rear)
March 1978 on (except 1300 base) ... Gas-pressurized, telescopic, double-acting

Torque wrench setting
	lbf ft	kgf m
Wheel nuts/bolts — conical face	63 to 85	8.7 to 11.7
Wheel nuts/bolts — flat face	85 to 103	11.7 to 14.2

3 Carburation

Variable venturi (VV) carburettor — description
1 As from November 1979 (build code WM) the variable venturi (VV) carburettor is fitted in place of the Motorcraft fixed venturi carburettor. The new carburettor gives increased performance and improved fuel consumption largely due to better fuel atomisation and air/fuel control.
2 Unlike other types of variable venturi carburettors in current use, the Ford concept employs an air valve which is pivoted. The valve incorporates a tapered needle which is positioned in the main and secondary fuel jets to provide the correct amount of fuel in relation to the volume of air entering the engine.
3 The valve is actuated through a pivot and lever, by a diaphragm and spring which is in turn activated by the vacuum within the carburettor. This arrangment ensures that the correct air/fuel mixture is supplied to the engine during all operational conditions.
4 The carburettor incorporates a sonic idle system which operates in an identical manner to the system fitted to the fixed jet carburettor described in Chapter 3.
5 An indirect throttle control is fitted which is of the progressive cam and roller type. During initial throttle pedal movement the throttle valve movement is small but, as the pedal approaches maximum travel, the throttle valve movement increases. The arrangement aids economy, and enables the same carburettor to be fitted to other models in the range.
6 Damping of the air valve to prevent a flat spot, is provided by a restrictor in the vacuum passage to the control diaphragm. When the throttle valve is opened quickly, the restrictor prevents the air valve responding simultaneously which would otherwise provide a weak mixture temporarily.
7 A diaphragm type accelerator pump is fitted, similar to that fitted to the fixed venturi carburettor.
8 The automatic choke is of the coolant sensitive, bi-metallic spring type, and incorporates its own variable needle jet. A vacuum choke pull-down system is employed whereby the choke is released under cruising conditions.
9 An anti-dieseling (anti-run-on) solenoid valve is fitted to cut off the idle mixture supply when the ignition is switched off. The solenoid plunger blocks off the sonic idle discharge tube.

Variable venture (VV) carburettor — adjustment
10 Run the engine until it reaches normal operating temperature.
11 Two adjustments are necessary for normal maintenance; these are idle speed adjustment and mixture setting. The idle speed adjustment procedure is identical to that described in Chapter 3, but requires a tachometer to determine the correct idling speed as given in the Specifications section of this Supplement.
12 The mixture adjustment screw is 'tamperproofed' and it is not normally necessary to remove the plug as the adjustment is preset during manufacture. However, the plug can be prised free to adjust the CO level. It is preferable to use an exhaust gas analyser, but if not available adjust the screw to achieve the smoothest idle. Turn the screw clockwise to weaken the mixture and anti-clockwise to enrich it. It may be necessary to readjust the idle speed.

Variable venturi (VV) carburettor — air filter element renewal
13 Unscrew the retaining screws and unclip the cover. If the main body comes free, disconnect the vacuum hoses and ducting.
14 Remove the element and discard it (photo).
15 Wipe clean the air cleaner body, then insert the new element.
16 Refit the body and/or cover in reverse order, making sure that the gasket is correctly positioned on the top of the carburettor.

Variable venturi (VV) carburettor — removal and refitting
17 Remove the air cleaner as described in paragraph 13 but do not unclip the cover.
18 Disconnect the wire to the anti-run-on solenoid.
19 Disconnect and plug the main fuel supply pipe.
20 Prise the clip from the carburettor end of the throttle cable, then compress the plastic clip to release the cable from the support (photos).
21 Pull off the distributor vacuum pipe at the carburettor.
22 With the cooling system cool, temporarily release the radiator filler cap. Disconnect and plug the two hoses from the automatic choke housing after identifying them for position.
23 Remove the two nuts and washers and withdraw the carburettor over the studs (photo). Remove the gasket from the inlet manifold.

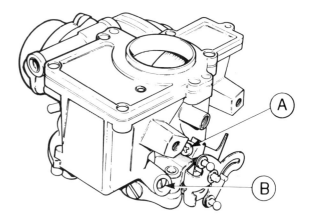

Fig. 13.1 Location of idle speed (A) and idle mixture (B) adjusting screws on the variable venturi carburettor (Sec 3)

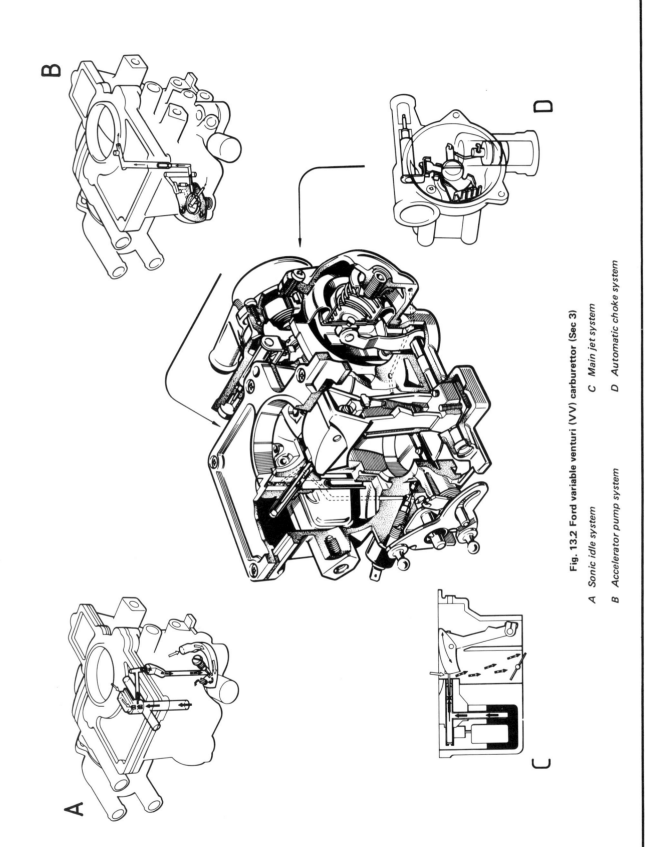

Fig. 13.2 Ford variable venturi (VV) carburettor (Sec 3)

A Sonic idle system
B Accelerator pump system
C Main jet system
D Automatic choke system

24 Refitting is a reversal of removal, but note the following additional points:

(a) Clean the carburettor and manifold mating faces and always use a new gasket
(b) Do not overtighten the retaining nuts
(c) Discard the crimped type fuel hose clamp if fitted, and fit a worm drive clip
(d) Adjust the idle speed and mixture setting as described in paragraphs 10 to 12 inclusive

Variable venturi (VV) carburettor – dismantling and reassembly

25 Clean the exterior of the carburettor with paraffin and wipe dry with a lint-free cloth.
26 Unscrew the seven screws and lift off the cover. Remove the gasket.
27 Drain the fuel from the float chamber. Prise the plug from the body and unscrew the metering needle from the air valve.
28 Remove the four screws and withdraw the main jet body from the top of the carburettor together with the gasket (photo).
29 Invert the carburettor and remove the accelerator pump outlet one-way valve ball and weight.
30 Extract the float pivot pin, and remove the float and needle valve needle (photo).
31 Remove the four screws and withdraw the control diaphragm housing, spring, and seat.
32 Pull back the diaphragm, prise out the C-clip, and lift out the diaphragm assembly (photo).
33 Remove the three screws and withdraw the accelerator pump housing, spring, and diaphragm.
34 Note the cover-to-body alignment marks, then remove the screws and withdraw the automatic choke cover, housing, and gasket (photo).
35 Unscrew the anti-run-on solenoid. The carburettor is now dismantled and the individual components can be cleaned and examined. Check each item for wear and damage, particularly linkages and moving parts. If the main and secondary jets are worn oval, renew the main jet body. Thoroughly check the diaphragm for splits or perishing. Make sure that the metering needle spring is correctly fitted. Obtain a kit of gaskets and diaphragms for fitting during reassembly.
36 Commence reassembly by locating the gasket face of the accelerator

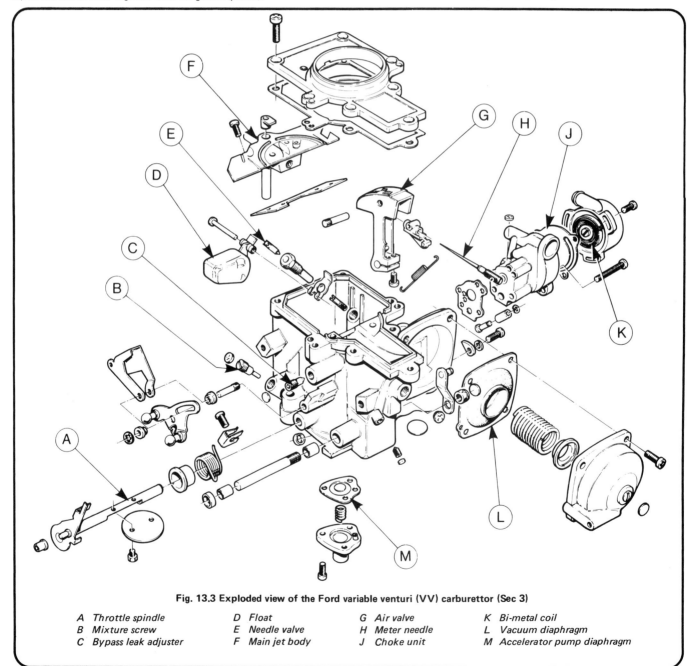

Fig. 13.3 Exploded view of the Ford variable venturi (VV) carburettor (Sec 3)

A Throttle spindle
B Mixture screw
C Bypass leak adjuster
D Float
E Needle valve
F Main jet body
G Air valve
H Meter needle
J Choke unit
K Bi-metal coil
L Vacuum diaphragm
M Accelerator pump diaphragm

3.14 Air cleaner element

3.20a Disconnecting the throttle inner cable from the carburettor

3.20b Depress the clips (arrowed) to remove the throttle cable

3.23 Removing the carburettor

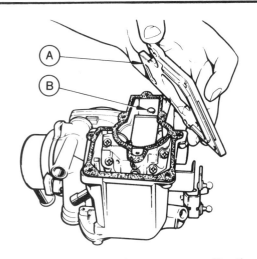
Fig. 13.4 Removing the carburettor top cover (Sec 3)

A Cover B Metering needle tamperproof plug

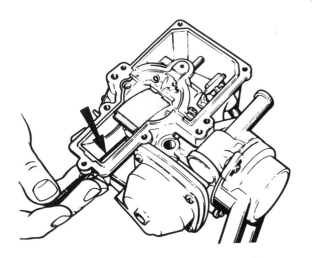
Fig. 13.5 Removing the metering needle (arrowed) (Sec 3)

3.28 Main jet body retaining screw locations (arrowed)

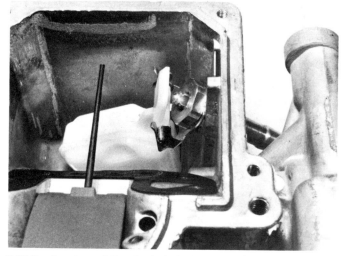

3.30 Needle valve and float (main jet body removed)

3.32 Diaphragm assembly retaining circlip location (arrowed)

3.34 Removing the automatic choke cover

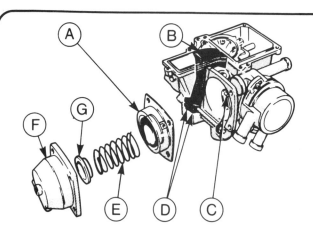

Fig. 13.6 Exploded view of the carburettor control diaphragm components (Sec 3)

  A Control diaphragm   E Spring
  B Air valve   F Cover
  C Pin   G Seat
  D Clamp screws

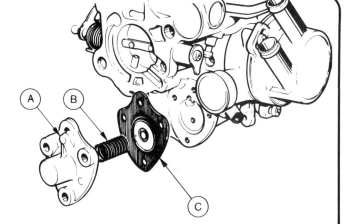

Fig. 13.7 Exploded view of the carburettor accelerator pump components (Sec 3)

  A Cover
  B Return spring
  C Diaphragm

Chapter 13 Supplement

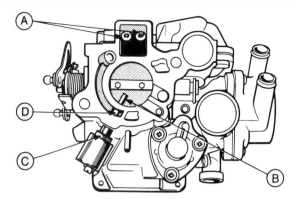

Fig. 13.8 Lower view of variable venturi carburettor (Sec 3)

- A Air valve adjusting screws
- B Sonic idle tube
- C Solenoid
- D Solenoid plunger

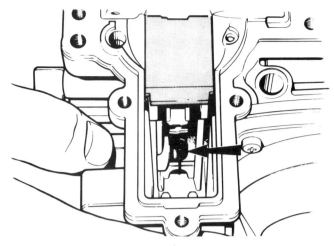

Fig. 13.9 Bias spring (arrowed) correctly fitted to air valve (Sec 3)

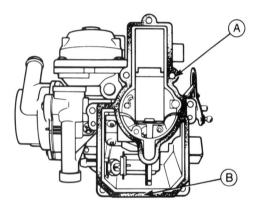

Fig. 13.10 Correct location of carburettor top cover gasket (Sec 3). Note particularly points A and B

3.47 Locate bi-metal coil in middle slot of choke lever (arrowed)

pump diaphragm on the housing, followed by the spring and cover. Insert and tighten the retaining screws.

37 Reconnect the control diaphragm assembly, ensuring that the double holes on one corner will align with those on the carburettor.

38 Locate the spring, seat, and cover over the diaphragm, then insert and tighten the retaining screws evenly in a diagonal sequence.

39 If the mixture adjustment screw tamperproof plug is removed, lightly tighten the screw then back it off three full turns.

40 Insert the needle valve needle with the spring loaded ball toward the float. Refit the float and pivot pin.

41 Insert the accelerator pump outlet one-way valve ball and weight into the discharge gallery.

42 Refit the main jet body with a new gasket, and tighten the retaining screws lightly in a diagonal sequence. Do not fully tighten the screws at this stage.

43 Slide the metering needle into the air valve and tighten it until the shoulder is aligned with the main jet body vertical face. Install the plug, using a liquid locking agent to secure it.

44 Open and close the air valve several times to centralise the main jet then, with the valve fully closed, tighten the main jet body screws in diagonal sequence.

45 Refit the top cover with a new gasket, and tighten the retaining screws evenly.

46 Refit the automatic choke housing using a new gasket, and tighten the retaining screws.

47 Engage the bi-metal coil in the cover with the middle slot in the choke lever, as the cover is fitted to the housing with a new gasket (photo). Align the cover-to-housing marks, and tighten the screws (see Fig. 13.11).

48 The carburettor is now ready to be fitted to the engine.

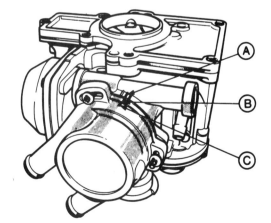

Fig. 13.11 Automatic choke alignment marks (Sec 3)

- A Centre index
- B Cut reference mark
- C Cast raised mark (not to be used for alignment)

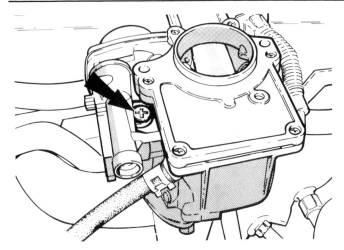

Fig. 13.12 Location of choke adjustment aperture plug (arrowed) (Sec 3)

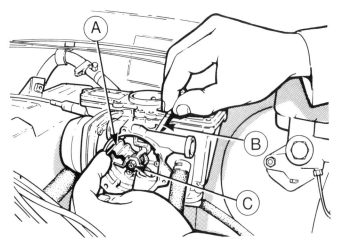

Fig. 13.13 Adjusting the automatic choke mixture (Sec 3)

A Choke lever
B Twist drill
C Shaft nut

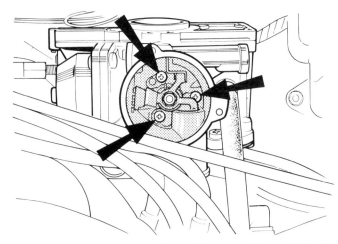

Fig. 13.14 Automatic choke housing retaining screws (arrowed) (Sec 3)

Variable venturi (VV) carburettor — automatic choke adjustment

Note: *The following procedure assumes that the carburettor is fitted to the engine although the adjustment can be made during reassembly of the carburettor on the bench.*

49 Remove the air cleaner.
50 Note the position of the alignment marks, then remove the screws and withdraw the automatic choke cover.
51 Prise the plug from the top of the carburettor behind the automatic choke housing. Look through the aperture and turn the automatic choke lever until the drilling in the shaft is in alignment. Insert a 3.4 mm (0.135 in) diameter twist drill fully into the drilling.
52 Loosen the nut on the end of the choke shaft. Turn the choke lever clockwise to its stop, then retighten the nut. This is the choke mixture adjustment.
53 Remove the twist drill, then bend the pull-down lever slightly downwards to ensure that it does not restrict the vacuum piston movement.
54 Look through the top aperture again and turn the lever to align the drilling. Insert a 4.5 mm (0.177 in) diameter twist drill fully into the drilling, then push the vacuum piston down to the bottom of its travel whilst holding the choke lever in the fully clockwise position. Note that there must be a clearance between the choke lever and pull-down lever, then bend back the pull-down lever so that it just touches the choke lever. This is the choke fast idle/pull-down adjustment.
55 Remove the twist drill and fit a new plug to the top of the carburettor.
56 Refit the automatic choke cover with the previously noted marks aligned (see Fig. 13.11), and the bi-metal spring engaged with the central slot. Tighten the screws evenly.
57 Refit the air cleaner.

Variable venturi (VV) carburettor automatic choke — removal and refitting

58 Remove the air cleaner assembly.
59 With the engine cold, release the pressure in the cooling system by temporarily removing the radiator cap. Disconnect and plug the hoses from the automatic choke cover after identifying them for position.
60 Note the cover-to-body alignment marks, then remove the screws and withdraw the cover.
61 Remove the three screws and withdraw the housing and gasket.
62 Refitting is a reversal of removal, but adjust the choke as described in paragraphs 51 to 55. Make sure that the bi-metal spring engages with the central slot in the choke lever. Top up the cooling system if necessary.

Variable venturi (VV) carburettor — float level check

63 If poor fuel consumption and difficult starting are experienced, the operating fuel level should be checked as follows.
64 Start the engine and run it at 3000 rpm for one minute, then switch off.
65 Remove the carburettor cover as previously described, and measure the distance from the fuel to the top of the casting. If the distance is not as given in the Specifications, remove the float and check it for leakage. Renew it if necessary.
66 Refit the carburettor cover.

Variable venturi (VV) carburettor — air valve adjustment

67 Two adjustments are possible; these are the maximum air valve movement, and the air valve-to-diaphragm lever setting.
68 To check the air valve movement, first remove the carburettor and withdraw the top cover as previously described.
69 With the air valve held fully open, measure the distance from the air valve to the main jet body. If this is not as given in the Specifications, remove the tamperproof plug and air valve stop screw, then coat the screw with a liquid locking agent and reinsert it to its correct position.
70 Fit a new plug and refit the carburettor.
71 To adjust the air valve-to-diaphragm lever setting, first remove the carburettor then dismantle the control diaphragm housing, spring, and diaphragm as previously described.
72 From under the carburettor loosen the two air valve screws.
73 Set the control diaphragm lever so that the distance from the housing mating face to the nearer edge of the diaphragm lever pin is as given in the Specifications.
74 Tighten the air valve screws and reassemble the control diaphragm housing components in reverse order.

Chapter 13 Supplement

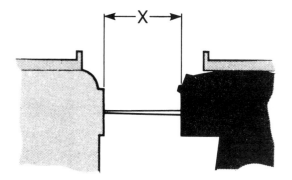

Fig. 13.15 Air valve maximum movement adjustment dimension X (Sec 3)

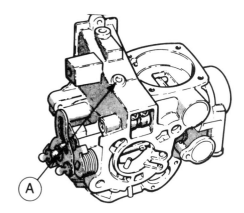

Fig. 13.16 Air valve maximum movement adjustment screw location (A) (Sec 3)

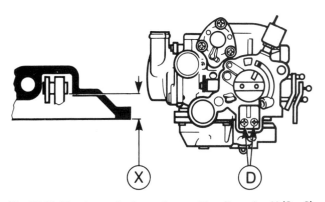

Fig. 13.17 Air valve-to-diaphragm lever setting dimension X (Sec 3)
D Air valve screws

75 Refit the carburettor as previously described.

4 Clutch

Release bearing – modification

1 As from October 1977 the clutch release bearing fitted is of the self-centering type as shown in Fig. 13.18. This type gives improved operation, particularly after the initial period of bedding-in.
2 If it is found that the original non-centering bearing is fitted during the removal of the clutch, it should always be removed and discarded, and the modified type fitted.

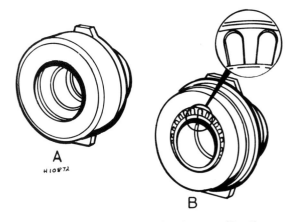

Fig. 13.18 Clutch release bearing types (Sec 4)

A Non self-centering   B Self-centering

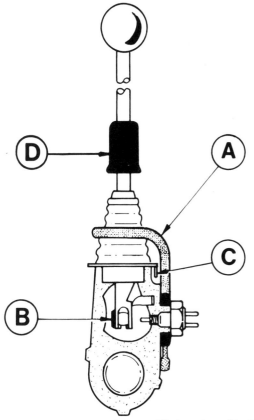

Fig. 13.19 Cross-section of the modified gear lever (Sec 5)

A Rubber strap   C Tab washer
B Damping bush   D Metal cup

5 Gearbox

Speedometer drivegear – modification

1 As from 1978 onwards, the speedometer drivegear shaft may be either of steel or plastic. Each type of shaft is fitted with its own type of seal, and it is important that the correct seal is fitted. Damage to the shaft may occur if the wrong seal is fitted; therefore, when renewing the seal, always check with the supplier.

Gear lever – modification

2 As from August 1980, an improved gear lever assembly is fitted as shown in Fig. 13.19. The modified lever can be identified by checking

the metal cup which fits over the rubber gaiter on the gear lever; if it does not incorporate inward protruding tabs, it is of the modified type.
3 If gear lever rattles are experienced, remove it as described in Chapter 6 and fit the modified components.
4 Lubricate the gear lever fork with grease, then refit it as described in Chapter 6. It will be necessary to select 3rd gear first.

6 Braking system

Rear brake modification — description
1 As from December 1974 (build code PG), a modified self-adjusting rear brake was progressively fitted, which is actuated by operation of the foot brake in contrast to the previous arrangement which was actuated by the handbrake.

Rear drum brake shoes (modified) — inspection and renewal
2 Remove the hub cap, loosen off the wheel nuts, then securely jack-up the car, and remove the roadwheel. Chock the front wheels and fully release the handbrake.
3 Lift off the drum. If necessary, lever out the handbrake lever abutment stop on the rear of the backplate to fully release the brake shoes.
4 Remove the small holding down springs from each shoe by turning the two small top washers through 90°.
5 Disengage the spring clip, and remove the clevis pin and washers securing the handbrake cable to the right-hand brake assembly at the handbrake actuating arm (on the left-hand brake assembly the handbrake rod must be disconnected from the lever).
6 The abutment stop must now be removed from the backplate. To remove the stop, remove the dust cover.
7 Carefully lever the brake shoes from their bottom guides and detach the lower return spring. Lever the leading shoe from the wheel cylinder slot and self-adjusting arm, disconnect the pull off spring and lift the shoe off the backplate.
8 Unlock the trailing shoe from the actuating lever, disconnect the pull off spring and lift the shoe off the backplate. Note that trailing/leading shoe identification is simplified by the fact that the lining is riveted to the trailing shoe, and bonded to the leading shoe.
9 The brake linings should be examined and must be renewed if they are so worn that the rivet heads are flush with the surface of the lining. Bonded linings must be renewed when the material has worn down to 1 mm (0.04 in) at its thinnest point.
10 Refitting the shoes is a direct reversal of the removal procedure, but great care must be taken to ensure that the return springs are correctly fitted and that the contact points shown in Fig. 13.22 are lightly greased.
11 Reset the self-adjusting mechanism by using a screwdriver to lever the serrated arm away from the wheel, at the same time pushing the serrated end of the arm towards the backplate to the limit of the serrations.
12 Adjust the brakes by applying the foot brake two or three times.

Rear drum brake wheel cylinder — removal and refitting
Note: *This procedure applies to the modified rear brake described in paragraph 1.*
13 Remove the brake drum and shoes as described in paragraphs 2 to 8.
14 Remove the self-adjusting mechanism. Where a fluid level warning switch is not fitted to the brake master cylinder reservoir, unscrew the cap and place a piece of polythene sheeting over the filler neck, then refit the cap.
15 Unscrew the hydraulic union(s) from the wheel cylinder and plug the pipe(s) to prevent leakage. Note that there are two unions on the right-hand side.
16 Unbolt and remove the wheel cylinder.
17 Refitting is a reversal of removal, but clean the mating faces of the wheel cylinder and backplate, and bleed the hydraulic system as described in Chapter 9.

Rear drum brake wheel cylinder — servicing
Note: *This procedure applies to the modified rear brake described in paragraph 1.*
18 Clean the exterior of the wheel cylinder.
19 Pull the rubber dust covers off the ends of the wheel cylinder.
20 Slide the piston assemblies, one from each end, out of the cylinder

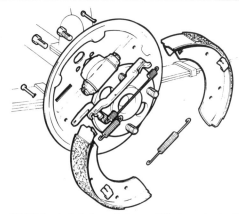

Fig. 13.20 Exploded view of the modified rear brake (Sec 6)

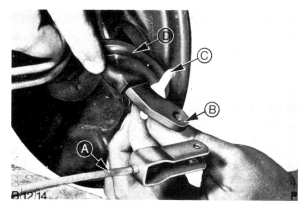

Fig. 13.21 Abutment stop location on the modified rear brake (Sec 6)

A Handbrake cable
B Lever
C Abutment stop
D Rubber boot

Fig. 13.22 Grease points on modified rear brake (Sec 6)

and withdraw the spring from the cylinder bore.
21 Prise the seals from the pistons and then examine the piston and bore for signs of scoring or excessive wear. If evident, renew the complete wheel cylinder. Discard the old seals and obtain a repair kit.
22 Dip the new inner seals in clean hydraulic fluid before fitting them to the pistons, and use the fingers only to manipulate them. Make sure that the tapered ends of the seals face outwards.
23 Insert the spring and piston assemblies into the cylinder, then fit the new dust covers.

Chapter 13 Supplement

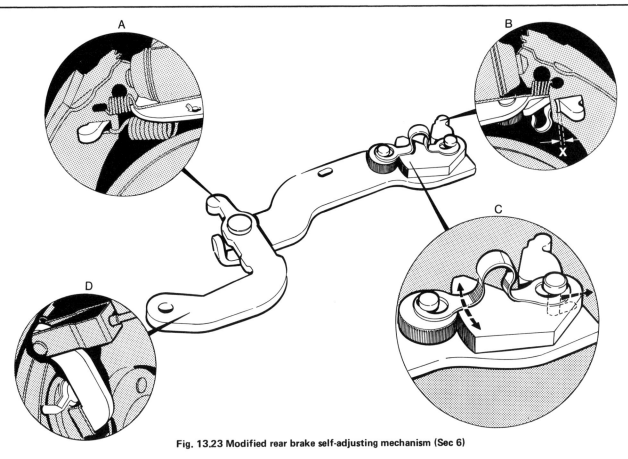

Fig. 13.23 Modified rear brake self-adjusting mechanism (Sec 6)

A Shoe end of handbrake actuating lever
B Shoe-to-drum clearance X (brakes released)
C Movement of serrated adjuster to reset mechanism
D Actuating lever and handbrake cable

Fig. 13.24 Exploded view of the abutment stop (Sec 6)

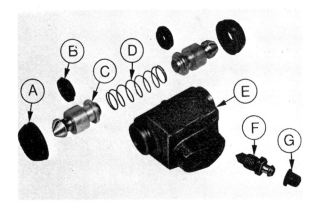

Fig. 13.25 Modified brake rear wheel cylinder components (Sec 6)

A Dust cover
B Seal
C Piston
D Spring
E Wheel cylinder
F Bleed nipple
G Cap

Handbrake — adjustment

Note: *This procedure applies to the modified rear brake described in paragraph 1.*

24 The handbrake is normally adjusted automatically by the self-adjusting mechanism, and this procedure will only therefore be necessary in the event of cable renewal or excessive stretch.
25 Chock the front wheels, then jack-up the rear of the car and support it on axle stands. Fully release the handbrake.
26 Check that the handbrake cable follows its correct run and is properly located in the guides. Make sure that there is no clearance between the lever abutment and the backplate on each rear brake.
27 Locate the cable adjuster on the underside of the floor pan. Using a screwdriver between the two shoulders, disengage the tapered adjusting nut from the keyed sleeve. Check that the keyed sleeve is engaged in the slot in the tunnel bracket.
28 Turn the adjusting nut *to remove all slack;* this is indicated by a *total* clearance of 4.0 to 5.0 mm (0.16 to 0.20 in) existing at the handbrake actuating lever abutment points. It is in order for zero clearance to exist on one side and the maximum total clearance on the other side, as both will equalise when the handbrake is operated.
29 Fully apply the handbrake, then release it to settle the mechanism. Check that the sum of the abutment clearances at each side totals between 3.0 and 4.5 mm (0.12 and 0.18 in); again it is in order for zero clearance to exist on one side.
30 The handbrake cable should not be tightened outside the above limits otherwise the self-adjusting mechanism will not operate correctly; with this system it is in order for the handbrake lever to travel up to 10 notches before the rear brakes are fully applied.

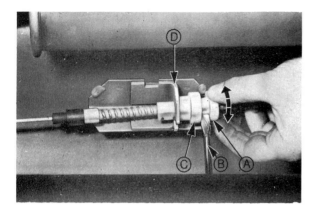

Fig. 13.26 Handbrake cable adjuster location on modified brake system (Sec 6)

- A Adjusting nut
- B Screwdriver holding nut and sleeve apart
- C Keyed sleeve
- D Bracket

Fig. 13.28 Brake pedal free height adjustment dimension X (Sec 6)

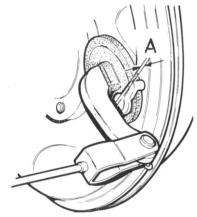

Fig. 13.27 Handbrake lever-to-abutment clearance A (Sec 6)

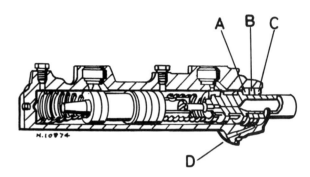

Fig. 13.29 Cut-away section of the Teves (ATE) brake master cylinder (Sec 6)

- A Secondary seal
- B Spacer
- C Vacuum seal
- D Drain hole

Brake pedal free height – checking

31 Chock the roadwheels and fully release the handbrake.
32 Using a pointed metal rod pressed through the carpet and sound deadener to the floor panel, measure the distance to the upper face of the brake pedal (see Fig. 13.28).
33 If it is not within the limits given in the Specifications, check the pedal linkage and bushes for wear, and renew them as necessary. Make sure that the floor panel and brake pedal have not been damaged. If still not correct, there may be a fault in the master cylinder or vacuum servo.

Fluid level warning indicator – description

34 As from October 1978 (build code UR), a fluid level warning indicator system is fitted, which consists of a float and switch assembly incorporated in the master cylinder reservoir cap; it is connected electrically to a warning light on the instrument panel.
35 The warning light is illuminated when the fluid level reaches the predetermined minimum level, as would occur if a leak was present in one of the hydraulic circuits.
36 The same warning light is also illuminated by the handbrake when on; therefore a check should be made that the handbrake is fully off before assuming that there is a hydraulic fluid leak.

Teves (ATE) master cylinder – description

37 Teves (ATE) master cylinders fitted to models manufactured from November 1978 on incorporate an additional seal in the mouth of the cylinder. The inner seal now retains the hydraulic fluid and the outer seal maintains vacuum in the vacuum servo.
38 The two seals are separated by a spacer which is vented to atmosphere by a small hole which also acts as a drain hole.

39 After a moderate period of use it is normal for a small accumulation of fluid to pass down the drain hole and stain the vacuum servo unit, but this does not indicate a faulty seal. Action should only be taken if droplets appear, in which case the master cylinder should be dismantled and overhauled, or renewed.

7 Electrical system

Electrical relays – description

1 As from February 1978 (build code UY), all electrical relays with the exception of the heated rear window relay have been turned through 180° to prevent water entry.
2 When renewing relays on models manufactured before this date, take the opportunity to turn all the following relays through 180°:

   (a) Wiper delay
   (b) Direction indicators
   (c) Fog lamps

Headlamps/sidelights – bulb renewal

3 For access to the headlamp assembly, first remove the retaining screw and take off the cover from the inner wing panel.
4 Turn the electrical connector/cap anti-clockwise and remove it.
5 To remove the headlamp bulb, disengage the retaining clip and swivel it upwards. Withdraw the bulb from the reflector and detach the cable plug.
6 When fitting the replacement headlamp bulb make sure that the glass envelope is not touched by the fingers; if necessary, use a clean cloth or tissue. Where the glass envelope has been touched by fingers, it can be cleaned with methylated spirit provided that the bulb has not been used.
7 To remove the sidelight bulb, pull the socket out of the reflector then remove the bulb by depressing and turning it.
8 Refitting the bulbs is the reverse of the removal procedure.

Chapter 13 Supplement

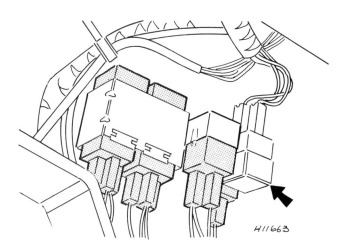

Fig. 13.30 Relay mounting from February 1978 on (heated rear window relay arrowed) (Sec 7)

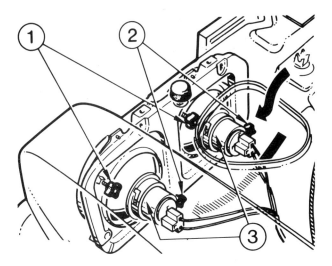

Fig. 13.31 Headlamp/sidelight bulb arrangement

1 Vertical alignment adjuster
2 Horizontal alignment adjuster
3 Electrical connector/cap
4 Retaining clip
5 Bulb (headlamp)
6 Socket
7 Bulb (sidelight)

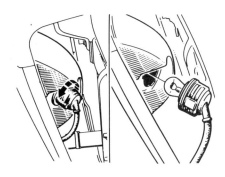

Fig. 13.32 Front direction indicator light bulb

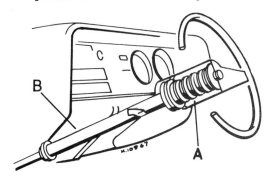

Fig. 13.33 Steering wheel collapsible can (A) and column tube (B) (Sec 8)

9 Headlamp bulbs can be adjusted vertically and horizontally by means of screws 1 and 2 (Fig. 13.31).

Front direction indicator lights — bulb renewal
10 Access to the bulbs is attained by reaching under the front bumper and turning the bulb socket anti-clockwise. Depress the bulb and turn it anti-clockwise to remove it from the socket.
11 Refitting is the reverse of the removal procedure.

8 Suspension and steering

Steering wheel — description
1 The steering wheel incorporates a collapsible can which is visible in photo 14.3, Chapter 11. Fig. 13.33 also illustrates the can.
2 The can is designed to collapse in the event of a front end impact, and it is important to realise that any damage to the can will impair its efficiency.
3 When removing the steering wheel, take care not to stretch or compress the convolutions of the can. If, on inspection, it is established that the can has been buckled or pressed from its original state, the steering wheel should be renewed at the earliest opportunity. Once distorted, it is not permissible to stretch or compress the can to its original dimension.

9 Bodywork and fittings

Window regulator knob — renewal
1 Models manufactured from September 1977 on are fitted with a window regulator knob which is designed to break off when subjected to a moderate impact. If this occurs, the knob can be renewed as follows.
2 Remove the window regulator handle as described in Chapter 12.
3 Using a screwdriver, rotate the remaining plastic retainer anti-clockwise by an eighth of a turn.
4 With the knob face down on a suitable tube of approximate internal diameter of 1.125 in (28 mm), press out the broken retainer and cap. Do not prise the cap out from the front.

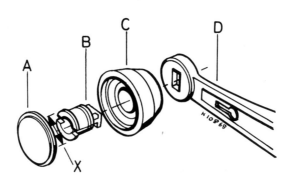

Fig. 13.34 Modified window regulator knob components (Sec 9)

A Cap
B Retainer
C Knob
D Arm

X Retainer diameter

5 The plastic retainer was manufactured in two diameters: 0.625 in (16 mm) up to late 1978, and 1.0 in (25.25 mm) from late 1978 on. If necessary, the larger diameter retainer can be used as a replacement for the earlier type.

6 Reassemble and refit the knob in reverse order, noting that the retainer must be turned an eighth of a turn clockwise to lock it.

Conversion factors

Length (distance)
Inches (in)	X	25.4	= Millimetres (mm)	X 0.039	= Inches (in)
Feet (ft)	X	0.305	= Metres (m)	X 3.281	= Feet (ft)
Miles	X	1.609	= Kilometres (km)	X 0.621	= Miles

Volume (capacity)
Cubic inches (cu in; in³)	X	16.387	= Cubic centimetres (cc; cm³)	X 0.061	= Cubic inches (cu in; in³)
Imperial pints (Imp pt)	X	0.568	= Litres (l)	X 1.76	= Imperial pints (Imp pt)
Imperial quarts (Imp qt)	X	1.137	= Litres (l)	X 0.88	= Imperial quarts (Imp qt)
Imperial quarts (Imp qt)	X	1.201	= US quarts (US qt)	X 0.833	= Imperial quarts (Imp qt)
US quarts (US qt)	X	0.946	= Litres (l)	X 1.057	= US quarts (US qt)
Imperial gallons (Imp gal)	X	4.546	= Litres (l)	X 0.22	= Imperial gallons (Imp gal)
Imperial gallons (Imp gal)	X	1.201	= US gallons (US gal)	X 0.833	= Imperial gallons (Imp gal)
US gallons (US gal)	X	3.785	= Litres (l)	X 0.264	= US gallons (US gal)

Mass (weight)
Ounces (oz)	X	28.35	= Grams (g)	X 0.035	= Ounces (oz)
Pounds (lb)	X	0.454	= Kilograms (kg)	X 2.205	= Pounds (lb)

Force
Ounces-force (ozf; oz)	X	0.278	= Newtons (N)	X 3.6	= Ounces-force (ozf; oz)
Pounds-force (lbf; lb)	X	4.448	= Newtons (N)	X 0.225	= Pounds-force (lbf; lb)
Newtons (N)	X	0.1	= Kilograms-force (kgf; kg)	X 9.81	= Newtons (N)

Pressure
Pounds-force per square inch (psi; lbf/in²; lb/in²)	X	0.070	= Kilograms-force per square centimetre (kgf/cm²; kg/cm²)	X 14.223	= Pounds-force per square inch (psi; lbf/in²; lb/in²)
Pounds-force per square inch (psi; lbf/in²; lb/in²)	X	0.068	= Atmospheres (atm)	X 14.696	= Pounds-force per square inch (psi; lbf/in²; lb/in²)
Pounds-force per square inch (psi; lbf/in²; lb/in²)	X	0.069	= Bars	X 14.5	= Pounds-force per square inch (psi; lbf/in²; lb/in²)
Pounds-force per square inch (psi; lbf/in²; lb/in²)	X	6.895	= Kilopascals (kPa)	X 0.145	= Pounds-force per square inch (psi; lbf/in²; lb/in²)
Kilopascals (kPa)	X	0.01	= Kilograms-force per square centimetre (kgf/cm²; kg/cm²)	X 98.1	= Kilopascals (kPa)

Torque (moment of force)
Pounds-force inches (lbf in; lb in)	X	1.152	= Kilograms-force centimetre (kgf cm; kg cm)	X 0.868	= Pounds-force inches (lbf in; lb in)
Pounds-force inches (lbf in; lb in)	X	0.113	= Newton metres (Nm)	X 8.85	= Pounds-force inches (lbf in; lb in)
Pounds-force inches (lbf in; lb in)	X	0.083	= Pounds-force feet (lbf ft; lb ft)	X 12	= Pounds-force inches (lbf in; lb in)
Pounds-force feet (lbf ft; lb ft)	X	0.138	= Kilograms-force metres (kgf m; kg m)	X 7.233	= Pounds-force feet (lbf ft; lb ft)
Pounds-force feet (lbf ft; lb ft)	X	1.356	= Newton metres (Nm)	X 0.738	= Pounds-force feet (lbf ft; lb ft)
Newton metres (Nm)	X	0.102	= Kilograms-force metres (kgf m; kg m)	X 9.804	= Newton metres (Nm)

Power
Horsepower (hp)	X	745.7	= Watts (W)	X 0.0013	= Horsepower (hp)

Velocity (speed)
Miles per hour (miles/hr; mph)	X	1.609	= Kilometres per hour (km/hr; kph)	X 0.621	= Miles per hour (miles/hr; mph)

*Fuel consumption**
Miles per gallon, Imperial (mpg)	X	0.354	= Kilometres per litre (km/l)	X 2.825	= Miles per gallon, Imperial (mpg)
Miles per gallon, US (mpg)	X	0.425	= Kilometres per litre (km/l)	X 2.352	= Miles per gallon, US (mpg)

Temperature

Degrees Fahrenheit (°F) = (°C × $\frac{9}{5}$) + 32

Degrees Celsius (Degrees Centigrade; °C) = (°F − 32) × $\frac{5}{9}$

*It is common practice to convert from miles per gallon (mpg) to litres/100 kilometres (l/100km), where mpg (Imperial) × l/100 km = 282 and mpg (US) × l/100 km = 235

Safety first!

Professional motor mechanics are trained in safe working procedures. However enthusiastic you may be about getting on with the job in hand, do take the time to ensure that your safety is not put at risk. A moment's lack of attention can result in an accident, as can failure to observe certain elementary precautions.

There will always be new ways of having accidents, and the following points do not pretend to be a comprehensive list of all dangers; they are intended rather to make you aware of the risks and to encourage a safety-conscious approach to all work you carry out on your vehicle.

Essential DOs and DON'Ts

DON'T rely on a single jack when working underneath the vehicle. Always use reliable additional means of support, such as axle stands, securely placed under a part of the vehicle that you know will not give way.

DON'T attempt to loosen or tighten high-torque nuts (e.g. wheel hub nuts) while the vehicle is on a jack; it may be pulled off.

DON'T start the engine without first ascertaining that the transmission is in neutral (or 'Park' where applicable) and the parking brake applied.

DON'T suddenly remove the filler cap from a hot cooling system – cover it with a cloth and release the pressure gradually first, or you may get scalded by escaping coolant.

DON'T attempt to drain oil until you are sure it has cooled sufficiently to avoid scalding you.

DON'T grasp any part of the engine, exhaust or catalytic converter without first ascertaining that it is sufficiently cool to avoid burning you.

DON'T allow brake fluid or antifreeze to contact vehicle paintwork.

DON'T syphon toxic liquids such as fuel, brake fluid or antifreeze by mouth, or allow them to remain on your skin.

DON'T inhale dust – it may be injurious to health (see *Asbestos* below).

DON'T allow any spilt oil or grease to remain on the floor – wipe it up straight away, before someone slips on it.

DON'T use ill-fitting spanners or other tools which may slip and cause injury.

DON'T attempt to lift a heavy component which may be beyond your capability – get assistance.

DON'T rush to finish a job, or take unverified short cuts.

DON'T allow children or animals in or around an unattended vehicle.

DO wear eye protection when using power tools such as drill, sander, bench grinder etc, and when working under the vehicle.

DO use a barrier cream on your hands prior to undertaking dirty jobs – it will protect your skin from infection as well as making the dirt easier to remove afterwards; but make sure your hands aren't left slippery. Note that long-term contact with used engine oil can be a health hazard.

DO keep loose clothing (cuffs, tie etc) and long hair well out of the way of moving mechanical parts.

DO remove rings, wristwatch etc, before working on the vehicle – especially the electrical system.

DO ensure that any lifting tackle used has a safe working load rating adequate for the job.

DO keep your work area tidy – it is only too easy to fall over articles left lying around.

DO get someone to check periodically that all is well, when working alone on the vehicle.

DO carry out work in a logical sequence and check that everything is correctly assembled and tightened afterwards.

DO remember that your vehicle's safety affects that of yourself and others. If in doubt on any point, get specialist advice.

IF, in spite of following these precautions, you are unfortunate enough to injure yourself, seek medical attention as soon as possible.

Asbestos

Certain friction, insulating, sealing, and other products – such as brake linings, brake bands, clutch linings, torque converters, gaskets, etc – contain asbestos. *Extreme care must be taken to avoid inhalation of dust from such products since it is hazardous to health.* If in doubt, assume that they *do* contain asbestos.

Fire

Remember at all times that petrol (gasoline) is highly flammable. Never smoke, or have any kind of naked flame around, when working on the vehicle. But the risk does not end there – a spark caused by an electrical short-circuit, by two metal surfaces contacting each other, by careless use of tools, or even by static electricity built up in your body under certain conditions, can ignite petrol vapour, which in a confined space is highly explosive.

Always disconnect the battery earth (ground) terminal before working on any part of the fuel or electrical system, and never risk spilling fuel on to a hot engine or exhaust.

It is recommended that a fire extinguisher of a type suitable for fuel and electrical fires is kept handy in the garage or workplace at all times. Never try to extinguish a fuel or electrical fire with water.

Fumes

Certain fumes are highly toxic and can quickly cause unconsciousness and even death if inhaled to any extent. Petrol (gasoline) vapour comes into this category, as do the vapours from certain solvents such as trichloroethylene. Any draining or pouring of such volatile fluids should be done in a well ventilated area.

When using cleaning fluids and solvents, read the instructions carefully. Never use materials from unmarked containers – they may give off poisonous vapours.

Never run the engine of a motor vehicle in an enclosed space such as a garage. Exhaust fumes contain carbon monoxide which is extremely poisonous; if you need to run the engine, always do so in the open air or at least have the rear of the vehicle outside the workplace.

If you are fortunate enough to have the use of an inspection pit, never drain or pour petrol, and never run the engine, while the vehicle is standing over it; the fumes, being heavier than air, will concentrate in the pit with possibly lethal results.

The battery

Never cause a spark, or allow a naked light, near the vehicle's battery. It will normally be giving off a certain amount of hydrogen gas, which is highly explosive.

Always disconnect the battery earth (ground) terminal before working on the fuel or electrical systems.

If possible, loosen the filler plugs or cover when charging the battery from an external source. Do not charge at an excessive rate or the battery may burst.

Take care when topping up and when carrying the battery. The acid electrolyte, even when diluted, is very corrosive and should not be allowed to contact the eyes or skin.

If you ever need to prepare electrolyte yourself, always add the acid slowly to the water, and never the other way round. Protect against splashes by wearing rubber gloves and goggles.

When jump starting a car using a booster battery, for negative earth (ground) vehicles, connect the jump leads in the following sequence: First connect one jump lead between the positive (+) terminals of the two batteries. Then connect the other jump lead first to the negative (–) terminal of the booster battery, and then to a good earthing (ground) point on the vehicle to be started, at least 18 in (45 cm) from the battery if possible. Ensure that hands and jump leads are clear of any moving parts, and that the two vehicles do not touch. Disconnect the leads in the reverse order.

Mains electricity

When using an electric power tool, inspection light etc, which works from the mains, always ensure that the appliance is correctly connected to its plug and that, where necessary, it is properly earthed (grounded). Do not use such appliances in damp conditions and, again, beware of creating a spark or applying excessive heat in the vicinity of fuel or fuel vapour.

Ignition HT voltage

A severe electric shock can result from touching certain parts of the ignition system, such as the HT leads, when the engine is running or being cranked, particularly if components are damp or the insulation is defective. Where an electronic ignition system is fitted, the HT voltage is much higher and could prove fatal.

Index

A

Accelerator
    cable — 52
    pedal — 53
Alternator
    brushes — 111
    fault diagnosis — 111
    general information — 109
    maintenance — 109
    removal and refitting — 111
    special procedures — 109
Air cleaner — 47
Antifreeze solution — 44

B

Battery
    charging — 109
    electrolyte replenishment — 109
    maintenance — 108
    removal and refitting — 108
Big-end bearings
    removal — 23
    renovation — 27
Bodywork
    centre console — 162
    description — 150
    doors — 154 to 157
    face level vents — 166
    instrument panel crash pad — 162
    load space trim panel — 160
    maintenance — 150, 151, 153
    opening rear quarter glass — 160
    rear quarter trim panel — 157
    repair
        major damage — 153
        minor damage — 151
    specifications — 150
    tailgate
        lock — 161
        removal and refitting — 161
        striker plate — 162
        window glass — 154
Bonnet — 160
Braking system
    bleeding the hydraulic system — 101
    description — 94
    drum brakes
        backplate — 99
        shoes — 98
        shoes (modified brake) — 176
        wheel cylinder — 99
        wheel cylinder (modified brake) — 176
    fault diagnosis — 105
    flexible hose — 101
    fluid level warning indicator — 178
    front disc brakes
        caliper — 95
        disc and hub — 97
        pads — 95
    handbrake
        adjustment (early models) — 101
        adjustment (modified brake) — 177
        adjustment (models with modified rear brakes) — 177
        cable and rod (later models) — 102
        cables (early models) — 102
        control lever — 103
    master cylinder
        description — Teves (ATE) — 178
        removal and refitting — 99
        servicing — 100
    pedal free height checking — 178
    pressure differential valve — 101
    rear brake modification — 176
    specifications — 94, 167
    torque wrench settings — 94
    vacuum servo unit — 103
Bumpers — 153

C

Cam followers *see* **Tappets**
Camshaft
    refitting — 33
    removal — 22
    renovation — 28
Carburation specifications — 46, 167
Carburettor description — 48, 168
Carburettor (motorcraft single venturi)
    accelerator pump adjustment — 51
    choke plate pull-down adjustment — 51
    cleaning — 49
    fast idling adjustment — 52
    float setting — 51
    removal and refitting — 49
    slow running adjustment — 52
Carburettor, variable venturi (VV)
    adjustment — 168
    air filter element renewal — 168
    air valve adjustment — 174
    automatic choke
        adjustment — 174
        removal and refitting — 174
    description — 168
    dismantling and reassembly — 170
    float level check — 174
    removal and refitting — 168
Circuit diagrams — 132 to 138
Clutch
    adjustment — 67
    cable — 70
    description — 67
    fault diagnosis — 70
    overhaul — 69
    pedal — 70
    refitting — 69
    release bearing
        modification — 175
        removal and refitting — 69
    removal — 69
    specifications — 67
    torque wrench settings — 67
Condenser — 59
Connecting rods
    refitting — 30
    removal — 23
    renewal — 23
    renovation — 29
Contact breaker points
    adjustment — 58
    removal and refitting — 59
Conversion factors — 181
Cooling system
    description — 40
    draining — 41
    fault diagnosis — 45
    filling — 41
    flushing — 41
    specifications — 40
    torque wrench settings — 40
Crankcase ventilation system servicing — 26
Crankshaft
    refitting — 30
    removal — 25
    renovation — 27

Index

Crankshaft pulley refitting — 34
Cylinder bores renovation — 28
Cylinder head
    decarbonising — 29
    dismantling — 18
    refitting — 34
    removal — 18

D

Decarbonising — 29
Distributor
    dismantling — 59
    reassembly — 62
    refitting — 36, 62
    removal — 59
Doors
    exterior handle — 156
    lock — 156
    rattles — 154
    remote control handle — 155
    trim panel — 155
    weatherstrips — 157
    window
        frame — 157
        frame mouldings — 157
        glass — 156
        regulating knob renewal — 179

E

Electrical system
    circuit diagrams — 132 to 138
    clock — 124
    description — 108
    door pillar switches — 125
    fault diagnosis — 130
    flasher unit — 124
    front direction indicator — 119, 179
    front light switches — 126
    fuses — 123
    handbrake warning light — 125
    hazard warning switch — 124
    headlamps — 117, 178
    heated rear window
        switch — 125
        warning light — 125
    instrument cluster — 124
    instrument voltage regulator — 124
    instrument/warning lamp bulbs — 125
    interior light — 125
    licence plate lamp — 119
    luggage compartment lamp — 126
    parking lamp — 119
    rear lamps — 119
    rear window washer and wiper switch — 125
    relays — 178
    reverse light switch — 126
    specifications — 106, 167
    speedometer cable — 124
    steering column
        lock — 125
        multi-function switch — 124
    torque wrench settings — 108
    windscreen washer (front) switch — 125
    windscreen wiper (front) switch — 126
Engine
    ancillary components removal — 18
    description — 17
    dismantling — 18
    fails to start — 64
    fault diagnosis — 37
    final assembly — 36
    front mountings — 26
    ignition switch — 125
    installation — 37
    misfires — 64
    operations possible with engine in car — 17
    operations requiring engine removal — 17
    reassembly — 30
    removal — 17
    renovation — 27
    specifications — 13
    start-up after overhaul — 37
    torque wrench settings — 16
Exhaust system
    description — 53
    muffler unit — 55
    pipe unit — 55
    resonator unit — 55

F

Fanbelt — 44
Flywheel
    refitting — 34
    removal — 24
Front suspension *see* Suspension, front
Fuel filler cap — 162
Fuel gauge sender unit — 48
Fuel pump — 47
Fuel system
    description — 47
    fault diagnosis — 56
    specifications — 46, 167
    torque wrench settings — 46
Fuel tank
    cleaning and repair — 48
    filler pipe — 48
    removal and refitting — 47
Fuses — 123

G

Gearbox
    description — 71
    dismantling — 72
    fault diagnosis — 83
    gear lever modification — 175
    input shaft — 77
    mainshaft — 77
    reassembly — 78
    removal and refitting — 71
    renovation — 75
    specifications — 71
    speedometer drive gear modification — 175
    torque wrench settings — 71
Gudgeon pins removal — 24

H

Heater
    controls — 164
    demister nozzles — 166
    dismantling and reassembly — 165
    fault diagnosis — 166
    removal and refitting — 164
    water valve (heavy duty heater) — 164
Hood — 160
Horn — 123
Hub bearings, front — 142

I

Ignition system
    description — 57

Index

fault diagnosis — 64
specifications — 57, 167
timing — 63

J

Jacking — 9

L

Lamps — 117, 119, 125, 126, 178, 179
Lubrication chart — 10
Lubrication system — 25

M

Main bearings
    removal — 25
    renovation — 27
Maintenance, routine — 7

O

Oil filter — 26
Oil pump
    overhaul — 26
    refitting — 34

P

Piston rings
    removal — 24
    renovation — 28
Pistons
    refitting — 30
    removal — 23
    renovation — 28
Propeller shaft
    description — 84
    fault diagnosis — 85
    removal and refitting — 85
Pushrods refitting — 35

R

Radiator — 41
Radiator grille — 153
Radio
    fitting — 126
    suppression of interference — 127
Rear axle
    description — 86
    differential — 88
    differential carrier — 88
    fault diagnosis — 93
    halfshaft — 87
    halfshaft bearing/oil seal — 88
    pinion oil seal — 92
    removal and refitting — 86
    specifications — 86
    torque wrench settings — 86
Rear suspension *see* Suspension, rear
Rocker arm/valve adjustment — 35
Rocker shaft
    dismantling — 29
    refitting — 35
    renovation — 29
Routine maintenance — 7

S

Safety first! — 182
Seatbelts — 163
Spare parts, buying — 6
Spark plugs — 63
Speedometer cable — 124
Starter motor
    Bosch pre-engaged overhaul — 116
    description — 113
    inertia
        overhaul — 113
        removal and refitting — 113
        testing on engine — 113
    Lucas pre-engaged overhaul — 115
    pre-engaged
        removal and refitting — 115
        testing on engine — 114
Starter ring gear — 29
Steering
    angles and front wheel alignment — 148
    column — 148
    description — 140
    fault diagnosis — 149
    gear
        adjustments — 147
        overhaul — 147
        removal and refitting — 145
    specifications — 139
    torque wrench settings — 140
    wheel — 179
Sump
    refitting — 34
    removal — 23
Suspension, front
    description — 140
    fault diagnosis — 149
    hub bearings — 140
    specifications — 139, 168
    stabiliser bar — 142
    torque wrench settings — 140
    track control arm — 143
Suspension, rear
    description — 140
    fault diagnosis — 149
    leaf spring — 144
    shock absorber — 143
    specifications — 139, 168
    stabiliser bar — 143
    torque wrench settings — 140
Switches — 124, 125, 126

T

Tailgate — 161, 162
Tape players
    fitting — 126
    suppression of interference — 127
Tappets
    refitting — 33
    renewal — 22
    renovation — 29
Temperature gauge — 44
Thermostat — 42
Timing chain tensioner
    refitting — 23
    removal — 25
Timing cover
    refitting — 33
    removal — 20
Timing sprockets and chain
    renewal — 20
    renovation — 29
Tools — 11

Towing — 9
Tyres
   general information — 149
   pressures — 139
   sizes — 139

U

Universal joints inspection — 85

V

Valve guides renovation — 29
Valves
   adjustment — 35
   refitting — 34
   renovation — 29

Vehicle identification numbers — 6

W

Water pump — 43
Wheels — 149
Windscreen — 153
Windscreen washer
   front — 121
   rear — 119
Windscreen wiper
   arms and blades — 123
   mechanism — 121
   motor
      bush — 123
      dismantling and reassembly — 123
   motor and linkage removal and refitting — 119

Printed by
J H Haynes & Co Ltd
Sparkford Nr Yeovil
Somerset BA22 7JJ England